MATHEMATICS RESEARCH DEVELOPMENTS

HILBERT SPACES

PROPERTIES AND APPLICATIONS

MATHEMATICS RESEARCH DEVELOPMENTS

Additional books and e-books in this series can be found on Nova's website under the Series tab.

MATHEMATICS RESEARCH DEVELOPMENTS

HILBERT SPACES

PROPERTIES AND APPLICATIONS

LE BIN HO
EDITOR

Copyright © 2020 by Nova Science Publishers, Inc.

All rights reserved. No part of this book may be reproduced, stored in a retrieval system or transmitted in any form or by any means: electronic, electrostatic, magnetic, tape, mechanical photocopying, recording or otherwise without the written permission of the Publisher.

We have partnered with Copyright Clearance Center to make it easy for you to obtain permissions to reuse content from this publication. Simply navigate to this publication's page on Nova's website and locate the "Get Permission" button below the title description. This button is linked directly to the title's permission page on copyright.com. Alternatively, you can visit copyright.com and search by title, ISBN, or ISSN.

For further questions about using the service on copyright.com, please contact:
Copyright Clearance Center
Phone: +1-(978) 750-8400   Fax: +1-(978) 750-4470   E-mail: info@copyright.com.

NOTICE TO THE READER

The Publisher has taken reasonable care in the preparation of this book, but makes no expressed or implied warranty of any kind and assumes no responsibility for any errors or omissions. No liability is assumed for incidental or consequential damages in connection with or arising out of information contained in this book. The Publisher shall not be liable for any special, consequential, or exemplary damages resulting, in whole or in part, from the readers' use of, or reliance upon, this material. Any parts of this book based on government reports are so indicated and copyright is claimed for those parts to the extent applicable to compilations of such works.

Independent verification should be sought for any data, advice or recommendations contained in this book. In addition, no responsibility is assumed by the Publisher for any injury and/or damage to persons or property arising from any methods, products, instructions, ideas or otherwise contained in this publication.

This publication is designed to provide accurate and authoritative information with regard to the subject matter covered herein. It is sold with the clear understanding that the Publisher is not engaged in rendering legal or any other professional services. If legal or any other expert assistance is required, the services of a competent person should be sought. FROM A DECLARATION OF PARTICIPANTS JOINTLY ADOPTED BY A COMMITTEE OF THE AMERICAN BAR ASSOCIATION AND A COMMITTEE OF PUBLISHERS.

Additional color graphics may be available in the e-book version of this book.

Library of Congress Cataloging-in-Publication Data

ISBN: 978-1-53616-633-0

Published by Nova Science Publishers, Inc. † New York

CONTENTS

Preface		vii
Chapter 1	Some Considerations on Orthogonality, Strict Separation Theorems and Applications in Hilbert Spaces *Manuel Alberto M. Ferreira*	1
Chapter 2	Solution Estimates for Autonomous Differential Equations in a Hilbert Space with Several Delays *Michael Gil'*	19
Chapter 3	Controllability of Quasi-Linear Evolution Differential System in a Separable Banach Space *Bheeman Radhakrishnan*	41
Chapter 4	Derivations of Operator Algebras on Hypercomplex Hilbert Spaces and Related Modules *S. V. Ludkowski*	61
Chapter 5	On Analytic Solutions of the Driven, 2-Photon and Two-Mode Quantum Rabi Models *Yao-Zhong Zhang*	123
Chapter 6	Hilbert Space of Model Hamiltonians *Medha Sharma*	143

Chapter 7	Enlarged Hilbert Spaces and Applications in Quantum Physics *Le Bin Ho*	**153**

About the Editor   **179**

Index   **181**

Related Nova Publications   **185**

PREFACE

This collective book presents selected topics in the modern research of Hilbert space. Throughout this book, various mathematical properties of the Hilbert space and extended Hilbert space are given, accompanied by reliable solutions and exciting applications to scientific and engineering problems. Together, the book presents to readers a picture of the modern theory of Hilbert space in its complexness and usefulness. The book is accessible for graduate students and could be served as a reference for scholars.

The book includes seven chapters.

Chapter 1 first provides some general viewpoints on convex sets, projections, and orthogonality in Hilbert spaces. Later, the Riesz representation theorem for continuous linear functionals is presented. Then, strict separation theorems of convex sets are also given where sufficient conditions for the separation are discussed detailed. This chapter also discuss the importance of the Riesz representation and strict separation theorems in the demonstration as well as the convex programing implementation of the Kuhn-Tucker theorem and the minimax theorem.

In chapter 2, the estimated solutions for autonomous differential equations in Hilbert spaces are discussed. The solutions are called "mild solutions," and such differential equations include partial-differential equations, integro-differential equations with or without time delay. First, the estimated solutions independent from delays are given, resulted in the stability conditions which also do not depend on delays. Then, the estimated solutions and the stability test dependent on delays are derived. Finally, the proposed solutions are applied to several interesting problems, including the coupled systems of differential-delay equations, the Hilbert-Schmidt Hermitian equations, the ordinary integro-differential equations, and the partial integro-differential equations: all with delays.

Coming along with the contents discussed in chapter 2, the mild-solution controllability of quasi-linear evolution of differential equations in separable Banach spaces is introduced in chapter 3. Therein, the sufficient conditions for controllability are proved. Various methods have been used to obtain the results; these include the Hausdorff measure of noncompactness, the fixed point approach, and a new calculation method.

Together, the mild solutions, the stability, and the controllability of various classes of differential equations in Hilbert spaces serve as efficient tools and motivate to study numerous scientific and engineering problems connected with diffusion and heat-flow in materials with memory, viscoelasticity, nonlinear behavior of elastic strings, and many other physical phenomena.

Chapter 4 is devoted to a discussion of hypercomplex Hilbert space, a generalization of complex Hilbert space, with the focus on the infinite-dimensional Cayley-Dickson algebras. Therein, homomorphisms of infinite-dimensional Cayley-Dickson algebras and its operator algebras on the hypercomplex Hilbert are investigated. Then, derivations and automorphisms are given. Related modules over octonions and Cayley-Dickson algebras are also considered. The corresponding cohomology theory is also developed to prove the existence of the nonassociative Cayley-Dickson algebras. The results of this chapter can be used for further studies on structures of operator algebras on hypercomplex Hilbert spaces, their derivations and cohomologies, and structure of hypercomplex Hilbert modules. It also is a potential candidate for mathematical coding theory and technical applications.

The three remaining chapters, chapter 5, 6, and 7, are devoted to applications of Hilbert spaces in quantum physics, an attractive and extensive topic of modern physics.

In chapter 5, analytic representations of solutions for the quantum wave function have been found by applying the Bogoliubov transformations through the introduction of the Bargmann-Hilbert spaces. The analytic solutions for the three cases of the driven Rabi model without Z_2 symmetry, the 2-photon quantum Rabi model, and the two-mode quantum Rabi model are given explicitly. These models serve as the basis for understanding matter-light interactions and have a variety of applications ranging from quantum optics to solid-state semiconductor systems, molecular physics, and the field of cavities and circuit quantum electrodynamics, and so on.

Chapter 6 is devoted to a discussion of reduced Hilbert spaces of model Hamiltonians. Therein, a Hubbard model has been examining to show a method for generating the basis states and the Hamiltonian matrix. To reduce the basis state, and hence the size of the Hilbert space, various symmetries have been used, including the z-component symmetry of spins and the number operator. This method of generation basis is applicable to other variants of the Hubbard model such as the Anderson impurity model.

In Chapter 7, an enlarged Hilbert space and its application in quantum physics have been discussed. The chapter first introduces the basic concepts of the enlarged Hilbert spaces for both cases of pure and mixed quantum states. The method for implementing the corresponding enlarged system in various physical platforms is also given. Later, the applications of the enlarged Hilbert space are discussed, including noncausal transformation problem, real quantum bits, and two-state vector formalism. Finally, the chapter is devoted to a connection between the enlarged Hilbert spaces and the quantum simulation.

With this rich collection of different topics, I believe this book will contribute significantly to the current development of science and technology.

Le Bin Ho
Osaka, Japan

In: Hilbert Spaces: Properties and Applications
Editor: Le Bin Ho
ISBN: 978-1-53616-633-0
© 2020 Nova Science Publishers, Inc.

Chapter 1

SOME CONSIDERATIONS ON ORTHOGONALITY, STRICT SEPARATION THEOREMS AND APPLICATIONS IN HILBERT SPACES

Manuel Alberto M. Ferreira[*]
Instituto Universitário de Lisboa (ISCTE-IUL),
Information Sciences, Technologies and
Architecture Research Center (ISTAR-IUL),
Business Research Unit (BRU-IUL) Lisboa, Portugal

Abstract

After presenting some structural notions on Hilbert spaces, which constitute fundamental support for this work, we approach the goals of the chapter. First, study about convex sets, projections, and orthogonality, where we approach the optimization problem in Hilbert spaces with some generality. Then the approach to Riesz representation theorem in this field, important in the rephrasing of the separation theorems. Then we give a look to the strict separation theorems as well as to the main results of convex programming: Kuhn-Tucker theorem and minimax theorem. These theorems are very important in the applications. Moreover, the presented strict separation theorems and the Riesz representation theorem have key importance in the demonstrations of Kuhn-Tucker and minimax theorems and respective corollaries.

---
[*] Corresponding Author's E-mail: manuel.ferreira@iscte.pt.

Keywords: Hilbert spaces, convex sets, projections, orthogonality, Riesz representation theorem, Kuhn-Tucker theorem, minimax theorem

1. INTRODUCTION

Definition 1.1.

1. A Hilbert space is a complex vector space with an inner product that, as metric space, is complete.

2. A Hilbert space is designated, usually, H or I. Remember that.

Definition 1.2. *An inner product in a complex vector space H is a sesquilinear Hermitian and strictly positive functional on H.*

Observation.

1. In real vector spaces, "sesquilinear Hermitian" must be replaced by "bilinear symmetric,"

2. The inner product of two vectors x and y belonging to H, in this order, is denoted as $\langle x|y \rangle$,

3. The norm of a vector x is given by $\|x\| = \sqrt{\langle x|x \rangle}$,

4. The distance between two elements x and y belonging to H is $d(x,y) = \|x - y\|$.

Proposition 1.1. *The norm, in a space with inner products, satisfies the parallelogram rule:*

$$\|x - y\|^2 + \|x + y\|^2 = 2\left(\|x\|^2 + \|y\|^2\right). \tag{1.1}$$

For more details on these concepts see, for instance, [1–5].

2. CONVEX SETS AND PROJECTIONS

Then, we enounce and demonstrate a theorem that is a result of existence and uniqueness, fundamental in optimization, see [6].

Theorem 2.1. *Every closed convex set in a Hilbert space has only one element with minimal norm.*

Dem. Call C, the closed convex set and $d = \inf \|x\|$, $x \in C$. Under the assumed conditions, it is possible to find a sequence $\|x_n\|$ in C, called minimizing sequence, such that $d = \lim_n \|x_n\|$. By the parallelogram rule, it is

$$\left\|\frac{x_n - x_m}{2}\right\|^2 = \frac{1}{2}\left(\|x_n\|^2 + \|x_m\|^2\right) - \left\|\frac{x_n}{2} + \frac{x_m}{2}\right\|^2.$$

Nevertheless, as the second parcel of the second member of this equality is the norm square of an element of C,

$$\left\|\frac{x_n - x_m}{2}\right\|^2 \leq \frac{1}{2}\left(\|x_n\|^2 + \|x_m\|^2\right) - d^2 \to 0,$$

and so x_n is a **Cauchy sequence**.

As C is closed and H is complete, the limit element z belongs to C. And, by the inequality $|\|x\| - \|y\|| \leq \|x - y\|$, it follows that $\|z\| = d$.

Now, suppose that z_1 and z_2 are two elements of C with norm d. So, again by the parallelogram rule, it has

$$\left\|\frac{1}{2}(z_1 - z_2)\right\|^2 = d^2 - \left\|\frac{z_1}{2} - \frac{z_2}{2}\right\|^2 \leq 0,$$

and then $z_1 = z_2$. □

Be now a closed convex set C in H, and an element x belonging to H. Noting that $x - C$ is a closed convex set, it results in the following Corollary of Theorem 2.1:

Corollary 2.1. *Be C a closed convex set in H. For every element x in H, there is only one element in C that is the closest of x; that is, there is only one element $z \in C$ such that*

$$\|x - z\| = \inf \|x - y\|, \, y \in C.$$

For the moment, there is a result of existence and uniqueness for the optimization problem. However, unhappily, the demonstration is not constructive. It is not said how to determine that unique element. However, it is possible a better characterization, through a variational inequality, as we point in the following result, see [7, 8]:

Theorem 2.2. *Be C a closed convex set in H. For every x belonging to H, z is the only element in C closest - in norm - of x if and only if*

$$\operatorname{Re}[\langle x-z|z-y\rangle] \geq 0, \, \forall y \in C. \tag{2.1}$$

Dem. Every characterization of this type comes through a variational argument. Suppose that z is the only element closest in C, granted by Corollary 2.1. So, for any θ, $0 \leq \theta \leq 1$, we have $(1-\theta)z + \theta y \in C$ since $y \in C$, as C is convex. So,

$$g(\theta) = \left\| x - \left((1-\theta)z + \theta y\right) \right\|^2, \tag{2.2}$$

is a function twice continuously differentiable of θ. More:

$$g'(\theta) = 2\operatorname{Re}[\langle x - \theta y - (1-\theta)z | z - y\rangle], \text{ and} \tag{2.3}$$
$$g''(\theta) = 2\operatorname{Re}[\langle z - y | z - y\rangle]. \tag{2.4}$$

Then, so that z is the minimizing element, it is evident that it has to be

$$g'(0) \geq 0 \Leftrightarrow \operatorname{Re}[\langle x-z|z-y\rangle] \geq 0.$$

Suppose now that (2.1) is fulfilled for a given element z of C. Therefore, building again $g(\theta)$ as in (2.2), Eq. (2.1) allows concluding that $g'(0)$ is non-negative and, owing to (2.4), $g''(0)$ is non-negative. So $g(0) \leq g(1)$ for any $y \in C$ that is

$$\|x-z\|^2 \leq \|x-y\|^2, \forall y \in C.$$

Therefore, it proofs that z is the minimizing element in C. As already seen, such an element is unique. $\square$

Observation.

1. It is interesting to interpret geometrically (2.1). Consider the set of elements h belonging to H, such that

$$\operatorname{Re}[\langle x-z|h\rangle] = c = \operatorname{Re}[\langle x-z|z\rangle].$$

2. Indeed, a hyperplane that contains z, whose normal is $x - z$, is a convex set C support plane in the sense that

$$\operatorname{Re}[\langle x - z | z \rangle] = c, \forall z \in C \qquad (2.5)$$
$$\operatorname{Re}[\langle x - z | y \rangle] \leq c, \forall y \in C. \qquad (2.6)$$

3. As

$$\operatorname{Re}[\langle x - z | z - y \rangle] \geq 0$$
$$\Leftrightarrow \operatorname{Re}[\langle x - z | z \rangle] - \operatorname{Re}[\langle x - z | y \rangle] \geq 0$$
$$\Leftrightarrow \operatorname{Re}[\langle x - z | z \rangle] \geq \operatorname{Re}[\langle x - z | y \rangle],$$

the point z is the support point. $\square$

Now it is pertinent to present the following Definitions, see [9]:

Definition 2.1. *Given any closed convex set C in H, the application of H in H, making to correspond to each x the closest element of x in C, is called projection over C and is designated $P_C(\cdot)$. $P_C(x)$ is said the projection of x over C.*

Observation. $P_C(\cdot)$ is not necessarily linear and lets C invariant. $\square$

Definition 2.2. *A cone is a set with the following property: $tx, t \geq 0$ belongs to it since x belongs.*

Observation.

1. A cone is not necessarily convex [1].

2. Note that C is a convex cone if whenever x_1 and x_2 belong to C also $t_1 x_1 + t_2 x_2$ belong to C for any $t_1, t_2 \geq 0$. $\square$

Then it follows a Corollary of Theorem 2.2:

Corollary 2.2. *Suppose that C is a closed convex cone. Be z the projection of x over C. Then*

$$\operatorname{Re}[\langle x - z | z \rangle] = 0 \text{ and } \operatorname{Re}[\langle x - z | y \rangle] \leq 0, \forall y \in C. \qquad (2.7)$$

In addition, if an element z of C satisfies these relations, it is the projection of x over C. $\square$

---
[1] It is enough to think in two straight lines passing through the origin.

Corollary 2.3. Be M a closed vector subspace. So, for each $x \in H$, there is only one element of M that is the closest of x, being the projection of x over M such that

$$\langle x - P_M(x) | m \rangle = 0, \forall m \in M. \tag{2.8}$$

In this case, $P_M(\cdot)$ is linear and called projection operator corresponding to M. □

3. Orthogonality and Orthonormal Basis

Following [10–12]:

Definition 3.1. Vector x is orthogonal to vector y if $\langle x | y \rangle = 0$.

Definition 3.2. The set S orthogonal complement in a Hilbert space is the set of the whole elements orthogonal to any element of S. Designate it $S^\perp$.

Proposition 3.1.

1. If $S \neq \emptyset$, $S^\perp$ is a closed vector subspace.

2. If M is a closed vector subspace

   a) $(M^\perp)^\perp = M$,

   b) After (2.8):

$$x = P_M(x) + (x - P_M(x)), \tag{3.1}$$

where $P_M(x) \in M$ and $(x - P_M(x)) \in M^\perp$.

Observation. In Eq. (3.1), it is patent an orthogonal decomposition of x. That is, x is decomposed in the sum of two elements orthogonal to each other. One belongs to the subspace M and the other belongs to its orthogonal complement. Such a decomposition is unique in the sense that if $x = z_1 + z_2$ where $z_1 \in M$ and $z_2 \in M^\perp$, it must be $z_1 = P_M(x)$ and $z_2 = x - P_M(x)$, since $(P_M(x) - z_1) + (x - (P_M(x) - z_2)) = 0$ and the elements between parenthesis are orthogonal. □

Definition 3.3. Call an orthonormal set if and only if any two of its elements are orthogonal to each other, and each element has norm 1.

Definition 3.4. *Be S a non-empty set of H. $L(S)$ designates the closure of the set of every S elements finite linear combinations.*

Definition 3.5. *An orthonormal set S is a basis of $L(S)$.*

Observation.

1. If S has a finite number of elements $x_i, i = 1, \cdots, n$, the subspace $L(S)$ is precisely the set of the whole elements of the form $\sum_{k=1}^{n} a_k x_k$. And, in this case, the projection operator corresponding to $L(S)$ is given by $P_{L(S)}(x) = \sum_{k=1}^{n} a_k x_k$ fulfilling the coefficients a_k the equation $\langle x - \sum_{j=1}^{n} a_j x_j | x_i \rangle = 0, i = 1, \cdots, n$ or:

$$\sum_{j=1}^{n} a_j \langle x_j | x_i \rangle = \langle x | x_i \rangle, i = 1, \cdots, n. \quad (3.2)$$

2. If the set $x_i, i = 1, \cdots, n$ is orthonormal, the projection has the simple form

$$P_{L(S)}(x) = \sum_{i=1}^{n} \langle x | x_i \rangle x_i, \quad (3.3)$$

and also

$$\left\| x \right\|^2 \geq \left\| P_{L(S)}(x) \right\|^2 = \sum_{i=1}^{n} \left| \langle x | x_i \rangle \right|^2 \text{ (Bessel's Inequality).}$$

3. Call now, S a sequence $\{x_i\}$ of elements $x_i, i = 1, \cdots, n$. S can be made orthonormal means that it is possible to determine an orthonormal basis for $L(S) : L(S) = L(O)$ being O orthonormal. Such a basis may be obtained through the well-known Gram-Schmidt method since not the whole $\{x_i\}$ are 0. □

With the whole generality:

Theorem 3.1. *Every non-trivial Hilbert space, that is not constituted exclusively by 0, has an orthonormal basis.*

Dem. It is possible to find orthonormal sets in the space, unless it is trivial. Introduce a partial ordination in the class of the orthonormal sets, through the inclusion relation:

Given two orthonormal sets A and B, $A < B$ if and only if $A \subset B$.

Be $\{A_\alpha\}$ a subclass totally ordered: a chain - maximal, that is: not strictly contained in another chain. The Hausdorff maximal chain theorem grants the existence of a maximal chain.

Be $A = \bigcup_\alpha A_\alpha$. A is orthonormal. Then, we show that $L(A)$, the subspace generated by A is, in fact, the whole Hilbert space.

Proceed by absurd. Suppose that $z \in H$ is not in $L(A)$. Call P the projection operator corresponding to $L(A)$. So $e = \dfrac{z - Pz}{\|z - Pz\|}$ is orthogonal to A and the family obtained postponing to the chain $\{A_\alpha\}$. The set $A \cup \{e\}$ violates the chain maximally. $\square$

Observation.

1. There may be, evidently, many sets as the set A referred in this demonstration, but it is demonstrated that all of them have the same cardinal.

2. An orthonormal basis may not be finite and the space is of infinite dimension. Moreover, it is not necessarily countable. However, it results from Bessel's inequality that, for every $x \in H$, only a countable number of $\langle x|e \rangle, e \in O$, may be different from zero. $\square$

4. RIESZ REPRESENTATION THEOREM

An important theorem about the representation of a continuous linear functional by elements of the space is the Riesz representation theorem, see again [11] and [13]:

Theorem 4.1 (Riesz representation). *Every continuous linear functional $f(\cdot)$ may be represented in the form $f(x) = \langle x|\tilde{q}\rangle$ where*

$$\tilde{q} = \dfrac{\overline{f(q)}}{\langle q|q \rangle} q,$$

and $\overline{f(q)}$ is the conjugate complex number of $f(q)$.

Dem. Begin noting that for every continuous linear functional $f(\cdot)$, the *Nucleus* of $f(\cdot)$ [2] is a closed vector subspace. If the functional under consideration is not the null functional, there is an element y such that $f(y) \neq 0$. Be z the projection of y over $\text{Nuc}(f)$ and make $q = y - z$. So, q is orthogonal to $\text{Nuc}(f)$ and $f(q) = f(y)$ and, in consequence, $f(q) \neq 0$. Then, for every $x \in H$, $x - \frac{f(x)}{f(q)} q$ belongs evidently to $\text{Nuc}(f)$. So, $x - \frac{f(x)}{f(q)} q$ is orthogonal to q and, in consequence,

$$\langle x|q\rangle - \frac{f(x)}{f(q)}\langle q|q\rangle = 0 \Leftrightarrow \langle x|q\rangle = \frac{f(x)}{f(q)}\langle q|q\rangle$$

that is: $f(x) = \left\langle x \left| \frac{\overline{f(q)}}{\langle q|q\rangle} q \right. \right\rangle$. $\square$

Observation. From the theorem, it results in $\|f\|_{H'} = \|\tilde{q}\|_H$, where the H dual space is H' [3]. $\square$

5. CONVEX SETS STRICT SEPARATION

Convex sets separation is very important in convex programming, which is a very potent mathematical instrument for operations research, management, and economics. See, for example, [14-16]. The target of this work is to present Theorem 5.1 that gives sufficient conditions for the strict separation of convex sets. First the following definitions:

Definition 5.1. *Two closed convex subsets A and B in a Hilbert space H are at a finite distance from each other if* $\inf_{x \in A,\ y \in B} \|x - y\| = d > 0$.

---

[2] The *Nucleus* of is designated $\text{Nuc}(f)$ and $\text{Nuc}(f) = \{x : f(x) = 0\}$.
[3] Consider a continuous linear functional f in a normed space E. It is called f norm, and designated $\|f\|$:

$$\|f\| = \sup_{\|x\| \leq 1} |f(x)|.$$

That is the supreme of the values assumed by $|f(x)|$ in the E unitary ball. The class of the continuous linear functionals, with the norm above defined, is a normed vector space, called the E dual space, designated E'. Of course, a Hilbert space is a normed space.

Definition 5.2. *Two closed convex subsets A and B in a Hilbert space H are strictly separated if, for some $v \in H$,*

$$\inf_{x \in A} \langle v|x \rangle > \sup_{y \in B} \langle v|y \rangle. \quad \square$$

Then it follows, see again [12].

Theorem 5.1 (Strict separation). *Two closed convex subsets A and B in a Hilbert space H at finite distance from each other can be strictly separated.*

Dem. Considering an $A - B$ complement interior point, taking its projection over the $A - B$ closure and calling it v, $\langle -v|v - q \rangle \geq 0, \forall q \in A - B$, by Theorem 2.2. So $\langle v|q \rangle \geq \langle v|v \rangle$ and $\langle v|x \rangle - \langle v|y \rangle \geq \langle v|v \rangle, x \in A, y \in B$ leading to $\inf_{x \in A} \langle v|x \rangle \geq \sup_{y \in B} \langle v|y \rangle. \quad \square$

It is also possible to show that:

Theorem 5.2. *Being H a finite-dimensional Hilbert space, if A and B are non-empty disjoint convex sets, they can always be separated.*

6. CONVEX PROGRAMMING

Now we outline a class of convex programming problems, at which we intend to minimize convex functionals subject to convex restrictions. Begin presenting a basic result that characterizes the minimum point of a convex functional subject to convex inequalities, see [17]. Note that it is not mandatory to impose any continuity conditions.

Theorem 6.1 (Kuhn-Tucker). *Be $f(x), f_i(x), i = 1, \cdots, n$, convex functionals defined in a convex subset C of a Hilbert space. Consider the problem $\min_{x \in C} f(x), sub. : f_i(x) \leq 0, i = 1, \cdots, n$. Be x_0 a point where the minimum, supposed finite, is reached. Suppose also that for each vector u in E_n, Euclidean space with dimension n, non-null and such that $u_k \geq 0$, there is a point x in C such that $\sum_1 u_k f_k(x) < 0$, designating u_k the components of u. So,*

1. There is a vector v, with non-negative components $\{v_k\}$, such that

$$\min_{x \in C} \left\{ f(x) + \sum_1^n v_k f_k(x) \right\} = f(x_0) + \sum_1^n v_k f_k(x_0) = f(x_0). \quad (6.1)$$

2. For every vector u in E_n with non-negative components, that is: belonging to the positive cone of E_n,

$$f(x) + \sum_1^n v_k f_k(x) \geq f(x_0) + \sum_1^n v_k f_k(x_0) \geq f(x_0) + \sum_1^n u_k f_k(x_0).$$

(6.2)

Corollary 6.1 (Lagrange duality). *In the conditions of Theorem 6.1,*

$$f(x_0) = \sup_{u \geq 0} \inf_{x \in C} f(x) + \sum_1^n u_k f_k(x). \square$$

Observation.

1. This Corollary is useful in supplying a process to determine the problem optimal solution.

2. If the whole v_k in expression (6.2) are positive, x_0 is a point in the border of the convex set defined by the restrictions.

3. If the whole v_k are zero, the inequalities do not influence the problem, that is: the minimum is equal to the one of the restrictions free problem. $\square$

Considering non-finite inequalities, see [18]:

Theorem 6.2 (Kuhn-Tucker in infinite dimension). *Be C a convex subset of a Hilbert space H and $f(x)$ a real convex functional defined in C. Be I a Hilbert space with a closed convex cone $\wp$, with non-empty interior, and $F(x)$ a convex transformation from H to I (convex in relation to the order introduced by cone $\wp$: if $x, y \in \wp, x \geq y$ if $x - y \in \wp$). Be x_0 an $f(x)$ minimizing in C subjected to the inequality $F(x) \leq 0$. Consider $\wp^* = \{x : \langle x|p \rangle \geq 0, \forall x \in \wp\}$ (dual cone). Admit that given any $u \in \wp^*$, it is possible to determine x in C such that $\langle u|F(x)\rangle < 0$. So, there is an element v in the dual cone $\wp^*$, such that for x in C*

$$f(x) + \langle v|F(x)\rangle \geq f(x_0) + \langle v|F(x_0)\rangle \geq f(x_0) + \langle u|F(x_0)\rangle,$$

being u any element of $\wp^*$. $\square$

Corollary 6.2 (Lagrange duality in infinite dimension).

$$f(x_0) = \sup_{v \in \wp^*} \inf_{x \in C} \Big(f(x) + \langle v | F(x) \rangle \Big)$$

in the conditions of Theorem 6.2. □

7. MINIMAX THEOREM

Although belonging to the field of convex programming, we make the option of giving privileged treatment to the Minimax Theorem, see [19, 20].

In a two players game with null sum, be $\Phi(x, y)$ a real function of two variables $x, y \in H$ and A and B convex sets in H. One of the players chooses strategies (points) in A in order to maximize $\Phi(x, y)$ (or minimize $-\Phi(x, y)$): it is the maximizing player. The other player chooses strategies (points) in B in order to minimize $\Phi(x, y)$ (or maximize $-\Phi(x, y)$); it is the minimizing player. The function $\Phi(x, y)$ is the payoff function. The function $\Phi(x_0, y_0)$ represents, simultaneously, the gain of the maximizing player and the loss of the minimizing player in a move at which they chose, respectively the strategies x_0 and y_0. So, the gain of one of the players is equal to the other's loss. That is why the game is a null sum game. A game in these conditions value is c if

$$\sup_{x \in A} \inf_{y \in B} \Phi(x, y) = c = \inf_{y \in B} \sup_{x \in A} \Phi(x, y). \tag{7.1}$$

If, for any (x_0, y_0), $\Phi(x_0, y_0) = c$, (x_0, y_0) is a pair of optimal strategies. There will be a saddle point if also

$$\Phi(x, y_0) \leq \Phi(x_0, y_0) \leq \Phi(x_0, y), \ x \in A, \ y \in B. \tag{7.2}$$

So, see again [6]:

Theorem 7.1 (minimax). *Consider A and B closed convex sets in H, being A bounded. Be a real functional defined for x in A and y in B fulfilling:*

1. $\Phi(x, (1-\theta)y_1 + \theta y_2) \leq (1-\theta)\Phi(x, y_1) + \theta\Phi(x, y_2)$ *for x in A and y_1, y_2 in B, $0 \leq \theta \leq 1$ (that is: $\Phi(x, y)$ is convex in y for each x.)*

2. $\Phi((1-\theta)x_1 + \theta x_2, y) \geq (1-\theta)\Phi(x_1, y) + \theta\Phi(x_2, y)$ *for y in B and x_1, x_2 in A, $0 \leq \theta \leq 1$ (that is: $\Phi(x, y)$ is convex in x for each y.)*

3. $\Phi(x, y)$ is continuous in x for each y.

So (7.2) holds, that is: the game has a value.

Dem. Beginning by the most trivial part of the demonstration:
$$\inf_{y \in B} \Phi(x, y) \leq \Phi(x, y) \leq \sup_{x \in A} \Phi(x, y),$$

and so
$$\sup_{x \in A} \inf_{y \in B} \Phi(x, y) \leq \inf_{y \in B} \sup_{x \in A} \Phi(x, y).$$

Then, as $\Phi(x, y)$ is concave and continuous in $x \in A$, A convex, closed and bounded, it follows that $\sup_{x \in A} \Phi(x, y) < \infty$.

Be $C = \inf_{y \in B} \sup_{x \in A} \Phi(x, y)$. Suppose now that there is $x_0 \in A$ such that $\Phi(x_0, y) \geq C$, for any y in B. In this case, $\inf_{y \in B} \Phi(x_0, y) \geq C$ or $\sup_{x \in A} \inf_{y \in B} \Phi(x, y) \geq C$ as it is appropriate. Then the existence of such an x_0 will be proved.

For any y in B, be $A_y = \{x \in A : \Phi(x, y) \geq C\}$. A_y is closed, limited, and convex. Suppose that, for a finite set $(y_1, y_2, \cdots, y_n)$, $\bigcap_{i=1}^{n} A_{y_i} = \emptyset$. Consider a transformation from A to E_n defined by

$$f(x) = \big(\Phi(x, y_1) - C, \Phi(x, y_2) - C, \cdots, \Phi(x, y_n) - C\big).$$

Call G the $f(A)$ convex hull closure. Be P the E_n closed positive cone. Now we show $P \cap G = \emptyset$: indeed, being $\Phi(x, y)$ concave in x, for any x_k in A, $k = 1, 2, \cdot, n$, $0 \leq \theta_k \leq 1$, $\sum_{k=1}^{n} \theta_k = 1$,

$$\sum_{k=1}^{n} \theta_k \big(\Phi(x_k, y) - C\big) \leq \Phi\left(\sum_{k=1}^{n} \theta_k x_k, y\right) - C.$$

Therefore, the convex extension of $f(A)$ does not intersect P.

Consider now a sequence x_n of elements of A, such that $f(x_n)$ converges for v, $v \in E_n$. As A is closed, limited and convex, it is possible to define a subsequence, designated x_m such that x_m converges weakly for an element of A (call it x_0). In addition, for any y_i as $\Phi(x, y_i)$ is concave in x,

$$\overline{\lim} \, \Phi(x_m, y_i) \leq \Phi(x_0, y_i), \text{ or } f(x_0) \geq \overline{\lim} \, f(x_m = v).$$

So $P \cap G = \emptyset$. Then, G and P may be strictly separated, and it is possible to find a vector in E_n with coordinates a_k, such that

$$\sup_{x \in A} \sum_{i=1}^{n} a_i \big(\Phi(x, y_i) - C\big) < \sum_{i=1}^{n} a_i e_i,$$

with the whole a_i greater or equal than zero.

Obviously, a_i cannot be simultaneously null. So dividing by $\sum_{i=1}^{n} a_i$ and taking in account the convexity of $\Phi(x, y)$ in y

$$\sup_{x \in A} \Phi(x, \bar{y}) - C < 0, \text{ where } \bar{y} = \frac{\sum_{k=1}^{n} a_k y_k}{\sum_{k=1}^{n} a_k}.$$

In addition, evidently, or $\bar{y} \in B$ or $\inf_{y \in B} \sup_{x \in A} \Phi(x, y) < C$. This contradicts the definition of C. So,

$$\bigcap_{i=1}^{n} A_{y_i} \neq \emptyset.$$

Indeed,

$$\bigcap_{y \in B} A_y \neq \emptyset.$$

as it will be seen in the sequence using that result and proceeding by absurd. Note that A_y is a closed and convex set and so it is also weakly closed. And being bounded, it is compact in the weak topology [4], as A. Calling G_y the complement of A_y, it results that G_y is open in the weak topology. So, if $\bigcap_{y \in B} A_y$ is empty, $\bigcap_{y \in B} G_y \supset H \supset A$. But, being A compact, a finite number of G_{y_i} is enough to cover A:

$$\bigcup_{i=1}^{n} G_{y_i} \supset A;$$

that is: $\bigcap_{i=1}^{n} A_i$ is in the complement of A and so it must be $\bigcap_{i=1}^{n} A_{y_i} = \emptyset$, leading to a contradiction. Then suppose that $x_0 \in \bigcap_{y \in B} A_y$. So, in fact x_0 satisfies $\Phi(x_0, y) \geq C$, as requested. □

Then it follows a Corollary of Theorem 7.1, obtained strengthening its hypothesis:

---

[4] See, for instance, [3].

Corollary 7.1. *Suppose that the functional* $\Phi(x,y)$ *defined in Theorem 7.1 is continuous in both variables, separately, and that B is limited. Therefore, there is an optimal pair of strategies, with the property of being a saddle point.*

Dem. It was already seen that exists x_0 such that

$$\Phi(x_0, y) \geq C, \tag{7.3}$$

for each y. As $\Phi(x_0, y)$ is continuous in y and B is limited

$$\inf_{y \in B} \Phi(x_0, y) = \Phi(x_0, y_0) \geq C, \tag{7.4}$$

for any y_0 in B [5]. But $\inf_{y \in B} \Phi(x_0, y) \leq \sup_{x \in A} \inf_{y \in B} \Phi(x, y) = C$ and, so

$$\Phi(x_0, y) = C. \tag{7.5}$$

The saddle point property follows immediately from (7.3), (7.4), and (7.5). □

To see more details about this approach of Minimax Theorem, see [21–26]. One last reference to Nash theorem, [27], which generalizes the Minimax Theorem:

Theorem 7.2 (Nash). *The mixed extension of every finite game has, at least, one strategic equilibrium.* □

Observation. Its demonstration demands, among other results, an important contribution of Kakutani theorem, see [28]. □

Acknowledgments

This work is financed by national funds through FCT - Fundação para a Ciência e Tecnologia, I.P., under the project UID/Multi/04466/2019. Furthermore, I would like to thank the Instituto Universitário de Lisboa and ISTAR-IUL for their support.

---

[5] A continuous convex functional in a Hilbert space has a minimum in any limited closed convex set.

CONCLUSION

Hilbert space is one of the mathematical fields more considered in the optimization problems fundamentals. Therefore, its structure and respective consequences deserve study and reflection. This was what we tried to do here in as simple ways as we could. It is always important to emphasize the fruitfulness of the results in part presented in convex programming, for instance in Kuhn-Tucker theorem and in the minimax theorem. It is never too much to point out the importance of strict separation theorems in achieving these results. Also to refer the importance of the Riesz representation theorem in the rephrasing of the separation theorems, key tools in functional optimization, here in strict separation theorems. Moreover, its direct contribution to getting the Lagrange duality results. Finally, to highlight Theorem 2.1, by its comprehensiveness, fundamental in optimization.

REFERENCES

[1] Aubin J. P. (1979), *Applied Functional Analysis*. John Wiley & Sons Inc., New York.

[2] Balakrishnan A. V. (1981), *Applied Functional Analysis*. Springer-Verlag New York Inc., New York.

[3] Kantorovich L. V. and Akilov G. P. (1982), *Functional Analysis*. Pergamon Press, Oxford.

[4] Kolmogorov and Fomin S. V. (1982), Elementos da Teoria das Funções e de Análise Funcional. Editora Mir.

[5] Royden H. L. (1968), *Real Analysis*. Mac Milan Publishing Co. Inc, New York.

[6] Ferreira M. A. M. (1986), Application of separation theorems in convex programming in Hilbert spaces. *Management Journal*, I, 41-44. (In Portuguese.)

[7] Ferreira M. A. M. and Filipe J. A. (2014), Convex sets strict separation in Hilbert spaces. *Applied Mathematical Sciences*, 8, 3155-3160.

[8] Ferreira M. A. M., Andrade M. and Matos M. C. (2010), Separation theorems in Hilbert spaces convex programming. *Journal of Mathematics and Technology*, 1, 20-27.

[9] Ferreira M. A. M. and Andrade M. (2011), Hahn-Banach theorem for normed spaces. *International Journal of Academic Research*, 3 (4, Part I), 13-16.

[10] Brézis H. (1983), *Analyse Fonctionelle (Théorie et Applications)*. Masson, Paris.

[11] Ferreira M. A. M. and Andrade M. (2011), Riesz representation theorem in Hilbert spaces separation theorems. *International Journal of Academic Research*, 3, 302-304.

[12] Ferreira M. A. M. and Andrade M. (2012), Separation of a vector space convex parts. *International Journal of Academic Research*, 4, 5-8.

[13] Ferreira M. A. M., Andrade M. and Filipe J. A. (2012), Weak convergence in Hilbert spaces. *International Journal of Academic Research*, 4, 34-36.

[14] Ferreira M. A. M. (2016), Optimization tools in management and finance. *Acta Scientiae et Intellectus*, 2, 45-59.

[15] Ferreira M. A. M. (2016), A topological approach to consumer theory. *Acta Scientiae et Intellectus*, 2, 15-19.

[16] Ferreira M. A. M. and Andrade M. (2011), Management optimization problems. *International Journal of Academic Research*, 3, 647-654.

[17] Ferreira M. A. M., Andrade M, Matos M. C., Filipe J. A. and Coelho M. (2012), Kuhn-Tucker's theorem - the fundamental result in convex programming applied to finance and economic sciences. *International Journal of Latest Trends in Finance & Economic Sciences*, 2, 111- 116.

[18] Ferreira M. A. M., Andrade M. and Filipe J. A. (2012), Kuhn-Tucker's theorem for inequalities in infinite dimension. *Journal of Mathematics and Technology*, 3, 57-60.

[19] von Neumann J. and Morgenstern O. (1971), *Theory of Games and Economic Behavior*. Princeton University Press, Princeton, New Jersey.

[20] von Neumann J. and Morgenstern O. (1967), *Theory of Games and Economic Behavior*. John Wiley & Sons Inc, New York.

[21] Ferreira M. A. M. (2015), The minimax theorem as Hahn-Banach theorem consequence. *Acta Scientiae et Intellectus*, 1, 58-66.

[22] Ferreira M. A. M., Andrade M, Matos M. C., Filipe J. A. and Coelho M. (2012), Minimax theorem and Nash equilibrium. *International Journal of Latest Trends in Finance & Economic Sciences*, 2, 36-40.

[23] Matos M. C. and Ferreira M. A. M. (2006), Game representation -code form. *Lecture Notes in Economics and Mathematical Systems*, 567, 321-334.

[24] Matos M. C, Ferreira M. A. M. and Andrade M. (2010), Code form game. *International Journal of Academic Research*, 2, 135-141.

[25] Matos M. C., Ferreira M. A. M., Filipe J. A. and Coelho M. (2010), Prisoner's dilemma: Cooperation or treason? *PJQM-Portuguese Journal of Quantitative Methods*, 1, 43- 52.

[26] Matos M. C, Ferreira M. A. M. and Filipe J. A. (2018), Let the games begin and go on. *International Journal of Business and Systems Research*. 12, 43-52.

[27] Nash J. (1951), Non-cooperative games. *Annals of Mathematics*, 54.

[28] Kakutani S. (1941), A generalization of Brouwer's fixed Point theorem. *Duke Mathematics Journal*, 8.

In: Hilbert Spaces: Properties and Applications
Editor: Le Bin Ho

ISBN: 978-1-53616-633-0
© 2020 Nova Science Publishers, Inc.

Chapter 2

SOLUTION ESTIMATES FOR AUTONOMOUS DIFFERENTIAL EQUATIONS IN A HILBERT SPACE WITH SEVERAL DELAYS

Michael Gil'*
Department of Mathematics, Ben Gurion University of the Negev,
Beer-Sheva, Israel

Abstract

Let $\mathcal{H}$ be a separable complex Hilbert space. The paper deals with equations of the type

$$\frac{dv(t)}{dt} = (Sv)(t) + \int_0^\eta B(\tau)v(t-\tau)d\mu(\tau) \ \ (t \geq 0; \ \eta < \infty),$$

where $\mu(\tau)$ is a bounded nondecreasing function, $B(\tau)$ is a piece-wise continuous function defined on $[0, \eta]$ whose values bounded are operators in $\mathcal{H}$, and S generates a strongly continuous semigroup on $\mathcal{H}$. Estimates for the norms of solutions to the considered equations are established. They give us explicit exponential stability conditions depending and independing on delays. The illustrative examples with the integro-differential equations with delays and coupled systems of differential-delay equations are presented.

Keywords: Hilbert space, functional differential equations, fundamental solutions, exponential stability

---

*Corresponding Author's E-mail: gilmi@bezeqint.net.

1. INTRODUCTION

Everywhere below, $\mathcal{H}$ is a separable complex Hilbert space with a scalar product $\langle\cdot|\cdot\rangle_\mathcal{H}$, the norm $\|\cdot\|_\mathcal{H} = \sqrt{\langle\cdot|\cdot\rangle_\mathcal{H}}$, and unit operator $I_\mathcal{H} = I$. By $\mathcal{B}(\mathcal{H})$, we denote the set of all bounded linear operators in $\mathcal{H}$.

The paper is devoted to the equation

$$\frac{dv(t)}{dt} = (Sv)(t) + \int_0^\eta B(\tau)v(t-\tau)d\mu(\tau) \quad (t \geq 0;\ \eta < \infty), \tag{1.1}$$

where $\mu(\tau)$ is a bounded nondecreasing function, $B(\tau)$ is a piece-wise continuous function defined on $[0, \eta]$ with values in $\mathcal{B}(\mathcal{H})$, and S generates a strongly continuous C_0-semigroup e^{St} on $\mathcal{H}$.

Such equations include partial differential-delay equations, integro-differential equations with delay and other traditional equations. Time-delay naturally appears in numerous equations. Recently, many papers were devoted to the stability of various classes of partial differential-delay equations. cf. [1–4] and references are given therein. Besides, mainly equations with one delay have been considered. The basic method for the stability analysis is the Lyapunov method extended to time-delay equations in a Hilbert space. By that method, many great results have been obtained. However, finding the Lyapunov-Krasovskij type functionals or solving the operator inequalities based on the corresponding operator equations are often connected with serious mathematical difficulties, especially regarding equations with many delays.

To the contrary, the exponential stability conditions presented in this chapter are formulated explicitly in terms of the spectra of the operator coefficients. This fact allows us to apply the well-known results from the spectral theory of operators. Our stability conditions are based on the norm estimates for solutions of the considered linear differential-delay equations. To the best of our knowledge, norm estimates for the solutions of abstract linear differential-delay equations have been established mainly in the case of one delay, cf. [5–8] and references therein, although they are very important for obtaining the stability tests and for the investigations of linear and non-linear perturbations of the considered equations (see for instance [9–13]). For examples, we consider the partial integro-differential equations with delays and coupled systems of parabolic differential-delay equations.

A few words about the contents: the paper consists of 9 sections. In Section 2, we introduce the main notations and prove the existence of solutions and their uniqueness. In Sections 3 and 4, we establish solution estimates independent of delays for equation (1.1). They give us the stability conditions, which are not dependent on delays. Section 5 is devoted to equations with bounded operators in a Hilbert space. Besides, solution estimates and a stability test dependent on delays are derived. Coupled systems of differential-delay equations are investigated in Section 6. Equations with operators having Hilbert-Schmidt Hermitian components are explored in Section 7. In Section 8, the results of Section 7 are applied to ordinary integro-differential equations. Section 9 deals with partial integro-differential equations.

2. NOTATIONS AND EXISTENCE OF SOLUTIONS

2.1. Notations

For a linear operator A, $\sigma(A)$ is the spectrum; $\text{Dom}(A)$ is the domain; $R_\lambda(A) := (A - I\lambda)^{-1}$ is the resolvent; $\lambda_k(A)$ ($k = 1, 2, \cdots$) are the eigenvalues taken with their multiplicities; A^* is the adjoint one; $\text{Im}[A] = (A - A^*)/2i$ is the imaginary Hermitian component; $\alpha(A) := \sup \text{Re}[\sigma(A)]$; $\text{dist}(A, \lambda) := \inf_{t \in \sigma(A)} |t - \lambda|$ is the distance between $\sigma(A)$ and a $\lambda \in \mathbb{C}$; and $\|A\| = \|A\|_{\mathcal{H}}$ means the operator norm if $A \in \mathcal{B}(\mathcal{H})$.

As usual, $L^p([a,b], \mathcal{H})$ ($1 \leq p < \infty$) is the space of $\mathcal{H}$-valued functions f defined on a real finite or infinite segment $[a, b]$ and equipped with the finite norm

$$\|f\|_{L^p(a,b)} = \|f\|_{L^p([a,b],\mathcal{H})} = \left[\int_a^b \|f(t)\|_{\mathcal{H}}^p dt\right]^{1/p}.$$

$C([a, b], \mathcal{H})$ is the space of $\mathcal{H}$-valued functions f defined and continuous on $[a, b]$, and equipped with the finite norm

$$\|f\|_{C(a,b)} = \|f\|_{C([a,b],\mathcal{H})} = \sup_{t \in [a,b]} \|f(t)\|_{\mathcal{H}}.$$

Also denoted by $Z(\eta, S)$ the set of functions ϕ defined on $[-\eta, 0]$ with values in $\text{Dom}(S)$ and such that $S\phi(t) \in C([-\eta, 0], \mathcal{H})$.

Similarly,

$$\|e^{St}\|_{L^1([a,b],\mathcal{H})} = \|e^{St}\|_{L^1(a,b)} := \int_a^b \|e^{St}\|_{\mathcal{H}} dt,$$

and
$$\|e^{St}\|_{C(R_+,\mathcal{H})} = \|e^{St}\|_{C(R_+)} := \sup_{t \in R_+} \|e^{St}\|_{\mathcal{H}}.$$

Here and below, $R_+ = [0, \infty)$, $R_\eta = [-\eta, \infty)$ and
$$\chi_0 := \int_0^\eta \|B(\tau)\|_{\mathcal{H}} d\mu(\tau).$$

2.2. Existence and Uniqueness of Solutions

In the sequel, it is assumed that $\mu(t)$ has a finite number of jumps, such that (1.1) can be written as

$$dv(t)/dt = (Sv)(t) + \sum_{k=0}^m B_k v(t-h_k) + \int_0^\eta \hat{B}(\tau) v(t-\tau) d\tau \quad (t \geq 0), \quad (2.1)$$

where $h_0 = 0, 0 < h_k = const \leq \eta$ $(k = 1, ..., m)$, $B_j \in \mathcal{B}(\mathcal{H})$ $(j = 0, ..., m)$, $\hat{B}(\tau)$ is a piece-wise continuous function defined on $[0, \eta]$ with values in $\mathcal{B}(\mathcal{H})$. Besides,

$$\chi_0 \leq \sum_{k=0}^m \|B_k\|_{\mathcal{H}} + \int_0^\eta \|\hat{B}(\tau)\|_{\mathcal{H}} d\tau.$$

Impose the initial condition
$$v(t) = \phi(t) \quad (-\eta \leq t \leq 0), \quad (2.2)$$

where $\phi \in Z(\eta, S)$ is given.

Let
$$(Ev)(t) := \sum_{k=0}^m B_k v(t-h_k) + \int_0^\eta \hat{B}(\tau) v(t-\tau) d\tau = \int_0^\eta B(\tau) v(t-\tau) d\mu(\tau) \quad (t \geq 0).$$

Then Eq. (2.1) can be rewritten as
$$dv(t)/dt = (Sv)(t) + (Ev)(t) \quad (t \geq 0). \quad (2.3)$$

Following Browder's terminology cf. [14, Chapter 5], a continuous function $v(t) : R_\eta \to \mathcal{H}$ satisfying

$$v(t) = e^{St}\phi(0) + \int_0^t e^{S(t-s)}(Ev)(s)ds \quad (t \geq 0), \quad (2.4)$$

we will call *a mild solution to (2.1, 2.2)*.
Obviously,

$$\int_0^t e^{S(t-s)}(Ev)(s)ds = \sum_{k=0}^m \int_0^t e^{S(t-s)} B_k v(s-h_k)ds + \int_0^\eta \int_0^t e^{S(t-s)} \hat{B}(\tau) v(s-\tau) ds\, d\tau$$

$$= \sum_{k=0}^m \left(\int_{h_k}^t e^{S(t-s)} B_k v(s-h_k)ds + \int_0^{h_k} e^{S(t-s)} \phi(t-h_k)ds \right)$$

$$+ \int_0^\eta \left(\int_\tau^t e^{S(t-s)} \hat{B}(\tau) v(s-\tau) ds + \int_0^\tau e^{S(t-s)} \hat{B}(\tau) \int_0^\tau \phi(s-\tau)ds \right) d\tau.$$

So Eq. (2.4) can be rewritten as

$$v(t) = f(t, E, \phi) + (Uv)(t), \tag{2.5}$$

where

$$f(t, E, \phi) = e^{St}\phi(0) + \sum_{k=0}^m \int_0^{h_k} e^{S(t-s)} B_k \phi(s-h_k)ds + \int_0^\eta \int_0^\tau e^{S(t-s)} \hat{B}(\tau) \phi(s-\tau) ds\, d\tau$$

and

$$(Uv)(t) := \sum_{k=0}^m \int_{h_k}^t e^{S(t-s)} B_k v(s-h_k)ds + \int_0^\eta \int_\tau^t e^{S(t-s)} \hat{B}(\tau) v(s-\tau) ds\, d\tau.$$

Besides,

$$\|Uv\|_{C([0,t],\mathcal{H})} \leq \chi_0 \|v\|_{C([0,t],\mathcal{H})} \int_0^t \|e^{Ss}\|_{\mathcal{H}} ds \quad (t > 0).$$

Clearly, U is a Volterra operator with a continuous kernel in $C([0,T],\mathcal{H})$ for any $T \in (0,\infty)$. So

$$(I-U)^{-1} = \sum_{k=0}^\infty U^k$$

and the series converges in the norm of $C([0,T],\mathcal{H})$. We thus have proved.

Lemma 2.1. *For any $\phi \in Z(\eta, S)$, the mild solution to problem (2.1, 2.2) exists, unique, and is defined by the equality*

$$v = (1-U)^{-1} f(\cdot, E, \phi). \tag{2.6}$$

A differentiable solution of problem (2.1, 2.2) is a function $v : R_\eta \to \mathrm{Dom}(S)$ having a strong derivative $\dot{v}(t) \in \mathcal{H}$ for each finite $t > 0$ and satisfying (2.1) and (2.2).

Let S be invertible and SES^{-1} be bounded. If $v(t)$ is a solution of the equation

$$(Sv)(t) = e^{St}S\phi(0) + \int_0^t e^{S(t-s)}(SES^{-1})(Sv)(s)ds \quad (t \geq 0), \qquad (2.7)$$

then it is also a solution of Eq. (2.4). Repeating our arguments of the proof of Lemma 2.1, we can assert that Eq. (2.7), with the initial function $S\phi(t)$ has a continuous solution. So the function $f_1(t) = (Ev)(t) \in \mathrm{Dom}(S)$ and $(Sf_1)(t)$ is continuous. By [15, Theorem I.6.5], $v(t)$ is differentiable and Eq. (2.4) is equivalent to

$$\dot{v} = Sv + f_1 \quad (v(0) = \phi(0)).$$

Hence due to the previous Lemma we arrive at:

Corollary 2.1. *If E maps $\mathrm{Dom}(S)$ into itself and SES^{-1} is bounded, then problem (2.1, 2.2) has a unique differentiable solution for any initial function from $Z(\eta, S)$.*

3. Solution Estimates Independent of Delays

In the sequel, the existence and uniqueness of solutions are assumed.

Definition 3.1. *Equation (2.1) is said to be exponentially stable if there are positive constants m_0, δ_0 independent of the initial function, such that*

$$\|v(t)\|_{\mathcal{H}} \leq \|\phi\|_{C([-\eta,0],\mathcal{H})} m_0 e^{-\delta_0 t} \quad (t \geq 0),$$

for any mild solution $v(t)$ of problem (2.1, 2.2).

Lemma 3.1. *Let*

$$\|e^{St}\|_{C(R_+,\mathcal{H})} < \infty \text{ and } \|e^{St}\|_{L^1(R_+,\mathcal{H})} < \frac{1}{\chi_0}. \qquad (3.1)$$

Then any mild solution $v(t)$ of problem (2.1, 2.2) is subject to the inequality

$$\|v\|_{C(R_+,\mathcal{H})} \leq \frac{1}{(1 - \|e^{St}\|_{L^1(R_+,\mathcal{H})}\chi_0)} \left[\|e^{St}\|_{C(R_+,\mathcal{H})} \|\phi(0)\|_{\mathcal{H}} + \chi_0\eta\|\phi\|_{C([-\eta,0],\mathcal{H})}\right]. \qquad (3.2)$$

Proof. From (2.5) it follows that

$$\|v\|_{C(0,t)} \leq \|f(\cdot, E, \phi)\|_{C(0,t)} + \|U\|_{C(0,t)} \|v\|_{C(0,t)} \quad (t > 0). \tag{3.3}$$

But

$$\|f(\cdot, E, \phi)\|_{C(0,t)} \leq \|e^{St}\|_{C(R_+)} \left[\|\phi(0)\|_{\mathcal{H}} + \chi_0 \eta \|\phi\|_{C(-\eta,0)}\right],$$

and

$$\|U\|_{C(0,t)} \leq \|e^{St}\|_{L^1(R_+,\mathcal{H})} \chi_0.$$

Consequently

$$\|v\|_{C(0,t)} \leq \|e^{St}\|_{L^1(R_+,\mathcal{H})} \chi_0 \|v\|_{C(0,t)} + \|e^{St}\|_{C(R_+)} \left[\|\phi(0)\|_{\mathcal{H}} + \chi_0 \eta \|\phi\|_{C(-\eta,0)}\right].$$

Now Eq. (3.1) implies

$$\|v\|_{C(0,t)} \leq \frac{1}{(1 - \|e^{St}\|_{L^1(R_+,\mathcal{H})} \chi_0)} \|e^{St}\|_{C(R_+)} \left[\|\phi(0)\|_{\mathcal{H}} + \chi_0 \eta \|\phi\|_{C(-\eta,0)}\right].$$

Hence, letting $t \to \infty$, we get the required result. □

Corollary 3.1. *Let conditions in (3.1) hold. Then (2.1) is exponentially stable.*

Indeed, substitute

$$v(t) = w(t)e^{-\epsilon t} \quad (\epsilon = const > 0), \tag{3.4}$$

into Eq. (2.1). Then

$$dw/dt = Sw + \epsilon w + \int_0^\eta B(\tau)e^{\epsilon \tau} w(t-\tau) d\mu(\tau) \quad (t \geq 0).$$

Taking ϵ sufficiently small, according to Eq. (3.2), we obtain

$$\|w\|_{C(R_+)} < \infty.$$

Making use of (3.4), we get

$$\|v(t)\|_{\mathcal{H}} < const\, e^{-\epsilon t} \quad (t \geq 0).$$

Hence we have the exponential stability. □

4. THE FUNDAMENTAL SOLUTION

Let $G(t)$ be a function defined on R_η, whose values are bounded operators in $\mathcal{H}$, such that
$$G(t) = 0 \ (t < 0), \ G(0) = I_\mathcal{H}. \tag{4.1}$$
Then it is called the (differentiable) Green function (or the (differentiable) fundamental solution) to Eq. (2.1) if $G(t)z$ is a differentiable solution of (2.1) for any $z \in \text{Dom}(S)$.

If together with conditions (4.1), $G(t)z$ is a mild solution of (2.1) for any $z \in \text{Dom}(S)$, then $G(t)$ will be called the mild Green function (or the mild fundamental solution) to Eq. (2.1).

Lemma 3.1 implies the below Corollary:

Corollary 4.1. *Let conditions in (3.1) hold. Then the (differentiable or mild) Green function of (2.1) satisfies the inequality*

$$\|G(t)\|_\mathcal{H} \le \frac{1}{(1 - \|e^{St}\|_{L^1(R_+,\mathcal{H})}\chi_0)} \|e^{St}\|_{C(R_+,\mathcal{H})} \ (t \ge 0).$$

5. ESTIMATES DEPENDENT ON DELAYS

In this section, we consider the equation with bounded operators:
$$\dot{w}(t) = (Ew)(t) \ (t \ge 0), \tag{5.1}$$
with
$$w(t) = \phi(t) \ (-\eta \le t \le 0), \tag{5.2}$$
where $\phi \in C([-\eta, 0], \mathcal{H})$. The definition of solutions to (5.1) is the same as above with $S = I_\mathcal{H}$. Since E is bounded, it can be easily checked that problem (5.1, 5.2) have a differentiable solution for any continuous $\phi(t)$. Denote by $G_E(t)$ the Green function of (5.1).

Lemma 5.1. *Let $G_E(t)$ be uniformly bounded on R_+. Then the solution $w(t)$ to problem (5.1, 5.2) satisfies the inequality*

$$\|w\|_{C(R_+,\mathcal{H})} \le c_1 \|\phi\|_{C([-\eta,0],\mathcal{H})},$$

where the constant c_1 does not depend on ϕ.

Proof. Put $u(t) = w(t) - \psi(t)$, where $\psi(t) = \phi(t)$ for $t < 0$ and $\psi(t) = \phi(0)e^{-at}$ ($a = const > 0$) for $t \geq 0$. Then Eq. (5.1) takes the form

$$\dot{u} = Eu + f,$$

where

$$f(t) = a\phi(0)e^{-at} + (E\psi)(t).$$

It is simple to see that

$$\|f(t)\|_{\mathcal{H}} \leq const \, \|\phi\|_{C([-\eta,0],\mathcal{H})} e^{-at} \quad (t \geq 0).$$

Since E is a bounded operator, it can be directly checked that

$$u(t) = \int_0^t G_E(t-s)f(s)ds.$$

Hence,

$$\sup_{t\geq 0} \|u(t)\|_{\mathcal{H}} \leq \sup_{t\geq 0} \|G_E(t)\|_{\mathcal{H}} \int_0^\infty \|f(s)\|ds.$$

This proves the Lemma. □

Rewrite (5.1) as

$$\dot{w}(t) = (Mw)(t) + (Tw)(t) \quad (t \geq 0), \tag{5.3}$$

where

$$M = \int_0^\eta B(\tau)d\mu(\tau) = \sum_{k=0}^m B_k + \int_0^\eta \hat{B}(\tau)d\tau \tag{5.4}$$

and

$$(Tw)(t) = \int_0^\eta B(\tau)\big(w(t-\tau) - w(t)\big)d\mu(\tau).$$

Assume that operator M is stable: $\alpha(M) = \sup \mathrm{Re}[\sigma(M)] < 0$. Since M is bounded, there are positive constants b and m_0, such that

$$\|e^{Mt}\|_{\mathcal{H}} \leq m_0 e^{-bt} \quad (t \geq 0).$$

So

$$\|e^{tM}\|_{L^1(R_+,\mathcal{H})} = \int_0^\infty \|e^{Ms}\|_{\mathcal{H}} ds < \infty.$$

Denote
$$\gamma(E) := \int_0^\eta \tau \|B(\tau)\|_{\mathcal{H}} d\mu(\tau).$$

Obviously,
$$\gamma(E) \le \sum_{k=1}^m h_k \|B_k\|_{\mathcal{H}} + \int_0^\eta \tau \|\hat{B}(\tau)\|_{\mathcal{H}} d\tau$$

and
$$\|E\|_{C([-\eta,t_0],\mathcal{H}) \to C([0,t_0],\mathcal{H})} \le \chi_0 \qquad (5.5)$$

for any $t_0 > 0$.

Lemma 5.2. *One has*
$$\|TG_E\|_{C(0,t_0)} \le \chi_0 \gamma(E) \|G_E\|_{C(0,t_0)} \quad (t_0 > 0).$$

Proof. Putting for simplicity $G_E(t) = v(t)$, we have $v(t) = 0$ ($t < 0$), $v(0) = I$, and due to (5.3):
$$v(t) = e^{Mt} + \int_0^t e^{M(t-s)}(Tv)(s)ds \quad (t \ge 0). \qquad (5.6)$$

But
$$v(s) - v(s-h) = \int_{s-h}^s \dot{v}(s_1) ds_1 \quad (h \ge 0).$$

and by (5.5)
$$\|\dot{v}\|_{C(0,t)} = \|Ev\|_{C(0,t)} \le \chi_0 \|v\|_{C(0,t)}.$$

Hence,
$$\|v(t-h) - v(t)\|_{C(0,t)} \le h\chi_0 \|v\|_{C(0,t)}.$$

So
$$\|Tv\|_{C(0,t)} \le \chi_0 \|v\|_{C(0,t)} \int_0^\eta \|B(\tau)\|_{\mathcal{H}} \tau \, d\mu(\tau) = \gamma(E)\chi_0 \|v\|_{C(0,t)},$$

as claimed. □

Lemma 5.3. Let M be a stable operator and

$$\gamma(E)\chi_0 \|e^{tM}\|_{L^1(R_+,\mathcal{H})} < 1. \tag{5.7}$$

Then

$$\|G_E\|_{C(R_+,\mathcal{H})} \le \frac{\|e^{Mt}\|_{C(R_+,\mathcal{H})}}{1 - \chi_0\gamma(E)\|e^{tM}\|_{L^1(R_+,\mathcal{H})}}.$$

Proof. Again, put $G_E(t) = v(t)$. Since

$$v(t) = e^{Mt} + \int_0^t e^{M(t-s)}(Tv)(s)ds \quad (t \ge 0),$$

we have

$$\|v\|_{C(0,t)} \le \|e^{Mt}\|_{C(R_+,\mathcal{H})} + \|e^{Mt}\|_{L^1(R_+,\mathcal{H})} \|Tv\|_{C(0,t)}.$$

By the previous Lemma

$$\|v\|_{C(0,t)} \le \|e^{Mt}\|_{C(R_+,\mathcal{H})} + \|e^{Mt}\|_{L^1(R_+,\mathcal{H})} \chi_0 \gamma(E) \|v\|_{C(0,t)}.$$

Condition (5.7) yields

$$\|v\|_{C(0,t)} \le \frac{\|e^{Mt}\|_{C(R_+,\mathcal{H})}}{1 - \|e^{Mt}\|_{L^1(R_+,\mathcal{H})} \chi_0 \gamma(E)}.$$

Hence the required result follows. □

Lemma 5.3 and a small perturbation of (5.1) as in (3.4) give us the following result:

Corollary 5.1. Under condition (5.7), equation (5.1) is exponentially stable.

6. COUPLED SYSTEMS OF DIFFERENTIAL-DELAY EQUATIONS

In this section, we consider Eq. (5.1) assuming that $\mathcal{H} = \mathbb{C}^n$-the n-dimensional Euclidean space ($n < \infty$) with the norm $\|\cdot\|_{\mathbb{C}^n}$. Then $\mathcal{B}(\mathcal{H}) = \mathbb{C}^{n \times n}$ is the set of all $n \times n$ matrices.

So, $B(\tau)$ is a bounded integrable on $[0, \eta]$ function with values in $\mathbb{C}^{n \times n}$. To apply Corollary 5.1, we need to estimate $\|e^{Mt}\|_{\mathbb{C}^n}$, where M is defined by (5.4). To this end, we introduce the quantity

$$g(A) = \left[N_2^2(A) - \sum_{k=1}^{n} |\lambda_k(A)|^2\right]^{1/2} \quad (A \in \mathbb{C}^{n \times n}),$$

where $\lambda_k(A), k = 1, ..., n$, are the eigenvalues of A, counted with their multiplicities; $N_2(A) = (\text{trace}\,(AA^*))^{1/2}$ is the Frobenius (Hilbert-Schmidt norm) of A. Here A^* is adjoint to A. The following relations are checked in [16, Section 3.1].

$$g^2(A) \leq N_2^2(A) - |\text{trace}\,A^2| \text{ and } g^2(A) \leq \frac{N_2^2(A - A^*)}{2} = 2N_2^2(\text{Im}[A]),$$

where $\text{Im}[A] = (A - A^*)/2i$. In addition, $g(e^{it}A + zI) = g(A)$ $(z \in \mathbb{C}, t \in \mathbb{R})$. If A is a normal matrix: $AA^* = A^*A$, then $g(A) = 0$.

If A_1 and A_2 are commuting matrices, then

$$g(A_1 + A_2) \leq g(A_1) + g(A_2).$$

By the inequality between geometric and arithmetic mean values,

$$\left(\frac{1}{n}\sum_{k=1}^{n}|\lambda_k(A)|^2\right)^n \geq \left(\prod_{k=1}^{n}|\lambda_k(A)|\right)^2.$$

So $g^2(A) \leq N_2^2(A) - n(\det A)^{2/n}$. Due to Example 3.2 from [16]

$$\|e^{tA}\|_{\mathbb{C}^n} \leq e^{\alpha(A)t} \sum_{k=0}^{n-1} \frac{t^k g^k(A)}{(k!)^{3/2}} \quad (t \geq 0),$$

where

$$\alpha(A) = \max_k \text{Re}[\lambda_k(A)].$$

If A is normal, then hence it follows the inequality

$$\|e^{tA}\|_{\mathbb{C}^n} \leq e^{\alpha(A)t} \quad (t \geq 0).$$

Assume that M is stable: $\alpha(M) < 0$ and

$$\int_0^\infty \|e^{Mt}\|_{\mathbb{C}^n} dt \leq \int_0^\infty e^{\alpha(M)t} \sum_{k=0}^{n-1} \frac{t^k g^k(M)}{(k!)^{3/2}} dt = \xi_n(M),$$

where
$$\xi_n(M) := \sum_{k=0}^{n-1} \frac{g^k(M)}{|\alpha(M)|^{k+1}\sqrt{k!}}.$$

If M is normal, then
$$\xi_n(M) = \frac{1}{|\alpha(M)|}.$$

In addition,
$$\sup_{t\geq 0} \|e^{Mt}\|_{C^n} \leq \zeta_n(M),$$

where
$$\zeta_n(M) = \sup_{t\geq 0} e^{\alpha(M)t} \sum_{k=0}^{n-1} \frac{t^k g^k(M)}{(k!)^{3/2}}.$$

In the case $\mathcal{H} = \mathbb{C}^n$, we have
$$\chi_0 = \int_0^{\eta} \|B(\tau)\|_{C^n} d\mu(\tau) \leq \sum_{k=1}^{m} \|B_k\|_{C^n} + \int_0^{\eta} \|\hat{B}(\tau)\|_{C^n} d\tau,$$

and
$$\gamma(E) = \int_0^{\eta} \tau \|B(\tau)\|_{C^n} d\mu(\tau) \leq \sum_{k=1}^{m} h_k \|B_k\|_{C^n} + \int_0^{\eta} \tau \|\hat{B}(\tau)\|_{C^n} d\tau.$$

Lemma 5.3 and Corollary 5.1 imply:

Corollary 6.1. *Let $\mathcal{H} = \mathbb{C}^n$. Let M be a stable $n \times n$-matrix defined by (5.4) and*
$$\gamma(E)\chi_0\xi_n(M) < 1. \tag{6.1}$$

Then
$$\|G_E(t)\|_{C(R_+)} \leq m_n(E, M),$$

where
$$m_n(E, M) := \frac{\zeta_n(M)}{1 - \chi_0\gamma(E)\xi_n(M)}.$$

Moreover, equation (5.1) is exponentially stable.

7. EQUATIONS WITH OPERATORS HAVING HILBERT-SCHMIDT HERMITIAN COMPONENTS

Recall that an operator A in a Hilbert space is called a Hilbert-Schmidt one, if

$$N_2(A) = \sqrt{\text{trace } AA^*} < \infty.$$

The ideal of all Hilbert-Schmidt operators is denoted by SN_2.

Again, consider equation (5.1). Define M by (5.4), assume that it is stable and

$$\text{Im}[M] = (M - M^*)/2i \in SN_2. \tag{7.1}$$

Numerous integro-differential operators satisfy this condition. Condition (7.1) is certainly fulfilled, if

$$B(\tau) - B^*(\tau) \in SN_2 \quad (0 \le \tau \le \eta).$$

Recall that $B(\tau)$ is bounded integrable on $[0, \eta]$ function with values in $\mathcal{B}(\mathcal{H})$.

Let $A - A^* \in SN_2$. Put

$$g_I(A) := \sqrt{2}\left[N_2^2(\text{Im}[A]) - \sum_{k=1}^{\infty}(\text{Im}[\lambda_k(A)])^2\right]^{1/2}.$$

Obviously,

$$g_I(A) \le \sqrt{2}N_2(\text{Im}[A]). \tag{7.2}$$

If A is normal, then $g_I(A) = 0$, cf. [16, Section 9.2]. Due to Example 10.1 from [16], for any bounded A with the property $A - A^* \in SN_2$ one has

$$\|e^{At}\|_{\mathcal{H}} \le \exp[\alpha(A)t] \sum_{k=0}^{\infty} \frac{g_I^k(A)t^k}{(k!)^{3/2}} \quad (t \ge 0), \tag{7.3}$$

where $\alpha(A) = \sup \text{Re}[\sigma(A)]$. If A is normal, then $\|e^{At}\| = e^{\alpha(A)t}$ $(t \ge 0)$. So inequality (7.3) is sharp.

If M is stable, according to (7.1) and (7.3) one has

$$\int_0^{\infty} \|e^{Mt}\|_{\mathcal{H}} dt \le \int_0^{\infty} e^{\alpha(M)t} \sum_{k=0}^{\infty} \frac{t^k g_I^k(M)}{(k!)^{3/2}} dt = \xi_{\infty}(M), \tag{7.4}$$

where
$$\xi_\infty(M) := \sum_{k=0}^{\infty} \frac{g_I^k(M)}{|\alpha(M)|^{k+1}\sqrt{k!}}.$$

If M is normal, then
$$\xi_\infty(M) = \frac{1}{|\alpha(M)|}.$$

Furthermore, one has
$$\sup_{t\geq 0} \|e^{Mt}\| \leq \zeta_\infty(M), \tag{7.5}$$

where
$$\zeta_\infty(M) := \sup_{t\geq 0} e^{\alpha(M)t} \sum_{k=0}^{\infty} \frac{t^k g_I^k(M)}{(k!)^{3/2}}.$$

Lemma 5.3 and Corollary 5.1 imply:

Corollary 7.1. *Let M be a stable operator defined by (5.4) and satisfying the conditions (7.1) and*
$$\gamma(E)\chi_0\xi_\infty(M) < 1. \tag{7.6}$$
Then the Green function $G_E(t)$ of (5.1) satisfies the inequality
$$\|G_E(t)\|_{C(R_+,\mathcal{E})} \leq m_\infty(M,E),$$
where
$$m_\infty(M,E) := \frac{\zeta_\infty(M)}{1 - \chi_0\gamma(E)\xi_\infty(M)}.$$
Moreover, equation (5.1) is exponentially stable.

8. ORDINARY INTEGRO-DIFFERENTIAL EQUATIONS

In this section, $\mathcal{H} = L^2(a,b)$ ($-\infty < a < b < \infty$) with the traditional scalar product. Consider the problem

$$\frac{\partial u(t,x)}{\partial t} = c(x)u(t,x) + \sum_{j=1}^{m} \int_a^b K_j(x,s)u(t-h_j,s)ds$$
$$+ \int_a^b \int_0^\eta \hat{K}(\tau,x,s)u(t-\tau,s)d\tau\,ds \quad (a \leq x \leq b, t \geq 0), \tag{8.1}$$

where $c(\cdot)$ is a real continuous function defined on $[a,b]$, $K_j(\cdot,\cdot)$ and $\hat{K}(\tau,\cdot,\cdot)$ ($j = 1,...,m$; $0 \le \tau \le \eta$) are complex in general functions, defined on $[a,b]^2$ and satisfying the conditions pointed below.

Equation (8.1) can be written as (5.1) with

$$(B_0 w)(x) = c(x)w(x), (B_j w)(x) = \int_a^b K_j(x,s)w(s)ds \ (1 \le j \le m)$$

and

$$(\hat{B}(\tau)w)(x) = \int_a^b \int_0^\eta \hat{K}(\tau,x,s)w(s)d\tau\,ds \ (w \in L^2(a,b)).$$

In addition, it is assumed that

$$N_2^2(B_j) = \int_a^b \int_a^b |K_j(x,s)|^2 ds\,dx < \infty \ (j = 1,...,m), \quad (8.2)$$

$$N_2^2(\hat{B}(\tau)) = \int_a^b \int_a^b |K_j(\tau,x,s)|^2 \, ds\,dx < \infty \text{ and } \int_0^\eta N_2(\hat{B}_2(\tau))d\tau < \infty. \quad (8.3)$$

Put

$$R(x,s) = \sum_{k=1}^m K_j(x,s) + \int_0^\eta \hat{K}(\tau,x,s)d\tau$$

and

$$(Ww)(x) = \int_a^b R(x,s)w(s)ds.$$

According to (5.4)

$$(Mw)(x) = c(x)w(x) + (Ww)(x).$$

Due to (8.2) and (8.3),

$$N_2(W) < \infty. \quad (8.4)$$

Since $c(x)$ is real, in view of (8.4), $M - M^*$ is a Hilbert-Schmidt operator and

$$g_I(M) \le \sqrt{2} N_2(\text{Im}[M]) \le \sqrt{2} N_2(W)$$

$$\le \sqrt{2} \sum_{j=1}^m N_2(B_j) + \sqrt{2} \int_0^\eta N_2(\hat{B}(\tau))d\tau.$$

Under consideration, we have

$$\chi_0 \leq \chi_\infty := \sup_x |c(x)| + \sum_{k=1}^m \|B_k\|_{L^2(a,b)} + \int_0^\eta \|\hat{B}(\tau)\|_{L^2(a,b)} d\tau$$

$$\leq \sup_x |c(x)| + \sum_{k=1}^m N_2(B_k) + \int_0^\eta N_2(\hat{B}(\tau)) d\tau,$$

and

$$\gamma(E) = \int_0^\eta \tau \|B(\tau)\|_{L^2(a,b)} d\mu(\tau)$$

$$\leq \sum_{k=1}^m h_k \|B_k\|_{L^2(a,b)} + \int_0^\eta \tau \|\hat{B}(\tau)\|_{L^2(a,b)} d\tau$$

$$\leq \sum_{k=1}^m h_k N_2(B_k) + \int_0^\eta \tau N_2(\hat{B}(\tau)) d\tau.$$

To apply Lemma 5.3 and Corollary 5.1, we need only to know the value $\alpha(M)$. The evaluation of $\alpha(M)$ is usually a difficult task. Let us point an estimate for $\alpha(M)$ in a practically important case when the considered integral operators are Volterra ones, i.e., (8.1) takes the form

$$\frac{\partial u(t,x)}{\partial t} = c(x)u(t,x) + \sum_{j=1}^m \int_a^x K_j(x,s)u(t-h_j,s)ds+$$

$$+ \int_a^x \int_0^\eta \hat{K}(\tau,x,s)u(t-\tau,s)d\tau\,ds \quad (a \leq x \leq b, t \geq 0). \quad (8.5)$$

In this case, W is a Volterra operator:

$$(Ww)(x) = \int_a^x R(x,s)w(s)ds,$$

and by Lemma 6.5.1 from [17] $\sqrt{2}N_2(\text{Im}[W]) = N_2(W)$. It is not hard to check that in this case

$$\alpha(M) = c_0 := \max_{x \in [a,b]} \sup c(x),$$

cf. [16, Lemma 8.6]. Assume that M is stable, i.e., $c_0 < 0$, and making use of (7.4), we get

$$\int_0^\infty \|e^{Mt}\|_{L^2(a,b)} dt \leq \int_0^\infty e^{c_0 t} \sum_{k=0}^\infty \frac{t^k N_2^k(W)}{(k!)^{3/2}} dt = \xi(W, c_0),$$

where

$$\xi(W, c_0) := \sum_{k=0}^\infty \frac{N_2^k(W)}{|c_0|^{k+1} \sqrt{k!}}.$$

In addition,

$$\sup_{t \geq 0} \|e^{Mt}\| \leq \zeta(W, c_0),$$

where

$$\zeta(W, c_0) = \sup_{t \geq 0} e^{c_0 t} \sum_{k=0}^\infty \frac{t^k N_2^k(W)}{(k!)^{3/2}}.$$

Now Lemma 5.3 and Corollary 5.1 imply:

Corollary 8.1. *Let $c_0 = \sup_x c(x) < 0$, the conditions (8.2), (8.3) and*

$$\gamma(E) \chi_\infty \xi(c_0, W) < 1$$

hold. Then the Green function $G_E(t, x)$ of equation (8.5) satisfies the inequality

$$\|G_E(t, \cdot)\|_{L^2(a,b)} \leq m(c_0, W) \quad (t \geq 0),$$

where

$$m(c_0, W) := \frac{\zeta(c_0, W)}{1 - \chi_\infty \gamma(E) \xi(c_0, W)}.$$

Moreover, equation (8.5) is exponentially stable.

9. A Partial Integro-Differential Equation

Put $\Omega = [0, 1] \times [a, b]$ and consider in $\mathcal{H} = L^2(\Omega)$, the equation

$$\frac{\partial u(t, x, y)}{\partial t} = \frac{\partial^2}{\partial y^2} u(t, x, y) + \int_a^x K(x, s) u(t, s, y) ds + c(x, y) u(t - h, x, y)$$

$$(a \leq x \leq b, 0 \leq y \leq 1, t \geq 0) \tag{9.1}$$

with the boundary condition
$$u(t, x, 0) = u(t, x, 1) = 0 \quad (t > 0; \; a \le x \le b), \tag{9.2}$$
where $c(\cdot, \cdot)$ is a scalar continuous function defined on Ω, $K(\cdot, \cdot)$ is a scalar function, defined on $[a, b]^2$ and satisfying the conditions pointed below. Put
$$\text{Dom}(S) = \{w \in L^2(\Omega) : \frac{\partial^2 w(x, y)}{\partial y^2} \in L^2(\Omega); \; w(x, 0) = w(x, 1) = 0; a \le x \le b\},$$
$$(S_1 w)(x, y) = \frac{\partial^2 w(x, y)}{\partial y^2}, \; (Vw) = \int_a^x K(x, s) w(s, y) ds \quad (w \in \text{Dom}(S))$$
and take $S = S_1 + V$. It is assumed that
$$N_2^2(V) = \int_a^b \int_a^x |K(x, s)|^2 ds \, dx < \infty.$$
Since V is a Volterra operator, its spectrum is zero and due to the above mentioned Lemma 6.5.1 from [17],
$$g_I(V) = \sqrt{2} N_2(\text{Im}[V]) = N_2(V).$$
By (7.3)
$$\|e^{Vt}\|_{\mathcal{H}} \le \sum_{k=0}^{\infty} \frac{N_2^k(V) t^k}{(k!)^{3/2}} \quad (t \ge 0).$$
In addition, S_1 is selfadjoint with the eigenvalues $-\pi^2 k^2$ ($k = 1, 2, ...$). So $\alpha(S_1) = -\pi^2$, and we can write $\|e^{S_1 t}\|_{\mathcal{H}} = e^{-\pi^2 t}$. Operators V and S_1 are commute. So $e^{St} = e^{S_1 t} e^{Vt}$ and therefore
$$\|e^{St}\|_{L^1(R_+)} \le \int_0^{\infty} e^{-\pi^2 t} \sum_{k=0}^{\infty} \frac{N_2^k(V) t^k}{(k!)^{3/2}} dt = \sum_{k=0}^{\infty} \frac{N_2^k(V)}{\pi^{2(k+1)} \sqrt{k!}}.$$
Under the consideration, we have $\chi_0 = \sup_{x,y} |c(x, y)|$. Making use of Corollary 3.1, we arrive at:

Corollary 9.1. *Let*
$$\sup_{x,y} |c(x, y)| \sum_{k=0}^{\infty} \frac{N_2^k(V)}{\pi^{2(k+1)} \sqrt{k!}} < 1.$$
Then equation (9.1) with boundary condition (9.2) is exponentially stable.

About integro-differential equations without delays see for example [18]. Some norm estimates for semigroups of non-selfadjoint operators can be found in [19, 20], and references therein.

REFERENCES

[1] Fridman E. and Orlov, Y. (2009), Exponential stability of linear distributed parameter systems with time-varying delays. *Automatica*, 45, 194-201.

[2] Iked A. K., Azuma T. and Uchida K. (2001), Infinite-dimensional LMI approach to analysis and synthesis for linear time-delay systems. *Special issue on advances in analysis and control of time-delay systems. Kybernetika (Prague)*, 37, 505-520.

[3] Luo Y. P. and Deng F. Q. (2006) LMI-based approach of robust control for uncertain distributed parameter control systems with time-delay. *Control Theory and Applications*, 23, 318-324.

[4] Wang L. and Wang Y. (2009) LMI-based approach of global exponential robust stability for a class of uncertain distributed parameter control systems with time-varying delays. *Journal of Vibration and Control*, 15, 1173-1185.

[5] Gil' M. I. (1998) On global stability of parabolic systems with delay. *Applicable Analysis*, 69, 57-71.

[6] Gil' M. I. (1998) On the generalized Wazewski and Lozinskii inequalities for semilinear abstract differential-delay equations. *Journal of Inequalities and Applications*, 2, 255-267.

[7] Gil' M. I. (2000) Stability of linear time-variant functional differential equations in a Hilbert space. *Funcialaj Ekvasioj*, 43, 31-38.

[8] Gil' M. I. (2002) Solution estimates for abstract nonlinear time-variant differential delay equations. *Journal of Mathematical Analysis and Applications*, 270, 51-65.

[9] Gil' M. I. (2002) Stability of solutions of nonlinear nonautonomous differential-delay equations in a Hilbert space. *Electronic Journal of Differential Equations*, 2002, 1-15.

[10] Gil' M. I. (2003) The generalized Aizerman - Myshkis problem for abstract differential-delay equations. *Nonlinear Analysis, TMA*, 55, 771-784.

[11] Gil' M. I. (2003) On the generalized Aizerman - Myshkis problem for retarded distributed parameter systems. *IMA Journal of Mathematical Control*, 20, 129-136.

[12] Gil' M. I. (2005) Stability of abstract nonlinear nonautonomous differential-delay equations with unbounded history-responsive operators. *Journal of Mathematical Analysis and Applications*, 308, 140-158.

[13] Gil' M. I. (2016) Explicit delay-dependent stability criteria for nonlinear distributed parameter systems, pages 291-315 in book *Advances and Applications in Nonlinear Control Systems, Studies in Computational Intelligence*, S. Vaidyanathan and C. Volos (eds.), 635, Springer, New York.

[14] Vrabie I. I. (1987) *Compactness Methods for Nonlinear Evolutions*. Pitman, New York.

[15] Krein S. G. (1971) *Linear Equations in a Banach Space*. Springer.

[16] Gil' M. I. (2018), *Operator Functions and Operator Equations*. World Scientific, New Jersey.

[17] Gil' M. I. (2003), *Operator Functions and Localization of Spectra*. N. Y. Springer.

[18] Gil' M. I. (2015), On stability of linear Barbashin type integro-differential equations. *Mathematical Problems in Engineering*, 2015, 5 pages.

[19] Gil' M. I. (2019), Semigroups of sums of two operators with small commutators. *Semigroup Forum*, 98, 22-30.

[20] Gil' M. I. (2019), Stability conditions for perturbed semigroups on a Hilbert space via commutators. *Communications in Advanced Mathematical Sciences*, 2, 129-134.

In: Hilbert Spaces: Properties and Applications
Editor: Le Bin Ho

ISBN: 978-1-53616-633-0
© 2020 Nova Science Publishers, Inc.

Chapter 3

CONTROLLABILITY OF QUASI-LINEAR EVOLUTION DIFFERENTIAL SYSTEM IN A SEPARABLE BANACH SPACE

Bheeman Radhakrishnan[*]
Department of Mathematics, PSG College of Technology,
Coimbatore, Tamil Nadu, India

Abstract

In this chapter, the sufficient conditions for controllability of quasi-linear evolution differential systems with non-local initial conditions in a separable Banach space have been proved. The results are obtained by using the Hausdorff measure of non-compactness, fixed point approach, and a new calculation method.

Keywords: controllability, quasi-linear differential system, measure of non-compactness, fixed point theorem

1. INTRODUCTION

A broad class of scientific and engineering problems is modeled by partial differential equations, integral equations or coupled ordinary and partial differential equations which can be described as differential equations in infinite-

---

[*]Corresponding Author's E-mail: radhakrishnanb1985@gmail.com.

dimensional spaces using semigroups. In general, functional differential equations or evolution equations serve as an abstract formulation of many partial integro-differential equations which arise in problems connected with heat-flow of materials with memory, viscoelasticity, and many other physical phenomena. So it becomes important to study the existence of solutions of such equations in infinite-dimensional spaces. Several authors have studied the problem represented by the evolution equations with bounded operators in Banach spaces.

It is well known that the systems described by partial differential equations can be expressed as abstract differential equations. Using the method of semigroups, various solutions of nonlinear and semi-linear evolution equations have been discussed by Pazy [1] and the non-local problem for the same equations has been first studied by Byszewskii [2]. Non-local Cauchy problem, namely, the differential equation with a non-local initial condition $x(t_0) + g(t_1, \ldots, t_p, x) = x_0$ $(0 \leq t_0 < t_1 < \ldots < t_p \leq t_0 + a$ and g is a given function) is one of the important topics in the study of analysis. Interest in such a problem stems mainly from the better effect of the non-local initial condition than the usual one in treating physical problems. There have appeared a lot of papers concerned with the existence of semi-linear evolution equations with non-local conditions [3, 4].

In today's rapidly progressing science and technology, the field of control theory is at the forefront of the creative interplay of mathematics, engineering and computer science. Drawing from these disciplines, control theory brings powerful theoretical results to bear upon advanced technologies. Roughly, the concept of controllability denotes the ability to move a system around its entire configuration space using only certain admissible manipulations. Controllability can steer a dynamic system from an arbitrary initial state to an arbitrary final state using the set of admissible controls. The complexity of modern systems, inaccuracies in output measurements and uncertainties about the system dynamics often make this problem extremely hard to solve.

As far as the controllability problems associated with finite-dimensional systems modelled by ODEs are concerned, this theory has been extensively studied during the last decades. In the finite-dimensional context, a system is controllable if and only if the algebraic Kalman rank condition is satisfied. According to this property, when a system is controllable for some time, it is controllable for all the time. But this is no longer true in the context of infinite-dimensional systems modelled by PDEs. In particular, in the frame work of wave equation, a model in which propagation occurs with finite velocity, in order that controllability properties to be true, the control time needs to be large enough so that the effect of the control may reach everywhere. When physical problems are simulated, the model often takes the form of semi-linear evolution equations.

Controllability of linear and nonlinear systems represented by ordinary differential equations in finite-dimensional spaces has been extensively investigated. Several papers have appeared on finite-dimensional controllability of linear systems [5] and infinite-dimensional systems in abstract spaces [6,7]. Of late, the controllability of nonlinear systems on finite-dimensional spaces by means of fixed point principles [8]. Several authors have extended the concept of controllability to infinite-dimensional spaces by applying semigroup theory [1, 9, 10]. Controllability of nonlinear systems with different types of nonlinearity has been studied by many authors with the help of fixed point principles [11]. Naito [12] discussed the controllability of nonlinear Volterra integro-differential systems, and in [13] he studied the controllability of semi-linear systems.

2. PRELIMINARIES

Consider a quasi-linear differential system with non-local conditions of the form

$$x'(t) = A(t, x(t))x(t) + Bu(t) + f(t, x(t)), \quad t \in J, \quad (2.1)$$
$$x(0) = g(x), \quad (2.2)$$

where the state variable $x(\cdot)$ takes values in the separable Banach space X with norm $\|\cdot\|$, $A(t, x(t))$ is the infinitesimal generator which generates a unique evolution system $\{S_x(t, s)\}_{0 \leq s \leq t \leq b}$ on a separable Banach space X, $g : C(J, X) \to X$ and $f : J \times X \to X$ are given mappings. The control function $u(\cdot)$ is given in $L^2(J, U)$, a Banach space of admissible control functions with U as a Banach space and $J = [0, b]$. In this chapter, we give conditions guaranteeing the controllability results for differential evolution system (2.1, 2.2) without assumptions on the compactness of f, g are condensing and the evolution system $\{S_x(t, s)\}$ is strongly continuous. The results obtained in this chapter are based on the new calculation method which employs the technique of measure of non-compactness.

Throughout this chapter, $(X, \|\cdot\|)$ be a real Banach space with zero element θ. Denote by

$$\mathbb{B}(x, r)$$

the closed ball in X centered at x and with radius r. The collections of all linear and bounded operators from X into itself will be denoted by $\mathcal{B}(X)$. If Y is a subset of X we write $\overline{Y}, ConvY$ in order to denote the closure and convex closure of Y, respectively.

Moreover, we denote by $\mathcal{F}_X$ the family of all nonempty and bounded subsets of X and by $\mathcal{G}_X$ its subfamily consisting of relatively compact sets.

Definition 2.1. *[14] A function $\chi : \mathcal{F}_X \to \mathbb{R}_+$ is said to be a regular measure of non-compactness if it satisfies the following conditions:*

1. *The family $\ker\chi = \{Y \in \mathcal{F}_X : \chi(Y) = 0\}$ is nonempty and $\ker\chi \subset \mathcal{G}_X$.*

2. $Y \subset Z \Rightarrow \chi(Y) \leq \chi(Z).$

3. $\chi(ConvY) = \chi(Y).$

4. $\chi(\lambda Y + (1-\lambda)Z) \leq \lambda\chi(Y) + (1-\lambda)\chi(Z), \text{ for } \lambda \in [0,1].$

5. *If $\{Y_n\}_{n=1}^{\infty}$ is a decreasing sequence of nonempty, bounded and closed subset of X such that $Y_{n+1} \subset Y_n$ $(n = 1, 2, ...)$ and if $\lim_{n \to \infty} \chi(Y_n) = 0$, then the intersection $Y_\infty = \bigcap_{n=1}^{\infty} Y_n$ is nonempty and compact in X.*

Remark. Let us notice that the intersection set Y_∞ described in axiom 5 is the kernel of measure of non-compactness χ. In fact, the inequality $\chi(Y_\infty) \leq \chi(Y_n)$, for $n = 1, 2, ...$ implies that $\chi(Y_\infty) = 0$. Hence $Y_\infty \in \ker\chi$. This property of the set Y_∞ will be important in our investigations.

Throughout this chapter, $\{A(t,s) : t \in \mathbb{R}\}$ is a family of closed linear operators defined on a common domain $\mathcal{D}$ which is dense in X, and we assume that the linear non-autonomous system

$$x'(t) = A(t,s)x(t), \quad s \leq t \leq b,$$
$$x(s) = x \in X, \tag{2.3}$$

has associated evolution family of operators $\{S_x(t,s) : 0 \leq s \leq t \leq b\}$. In the next definition, $\mathcal{L}(X)$ is a space of bounded linear operator from X into X endowed with the uniform convergence topology.

Definition 2.2. *Let $B \subset X$ and let $\{A(t,b)\}, (t,b) \in [0,T] \times B$ be a family of infinitesimal generator of a C_0 semigroup and Y is a Banach space, which is densely and continuously imbedded in X. If $x \in C([0,T] : X)$ has values in B then there is a unique evolution system $S_x(t,s) : 0 \leq s \leq t \leq b \subset \mathcal{L}(X)$ satisfying*

1. $\|S_x(t,s)\| \leq Me^{w(t-s)}$, $0 \leq s \leq t \leq T$;
2. $\frac{\partial^+}{\partial t} S_x(t,s)w|_{t=s} = A(s, x(s))w$;
3. $\frac{\partial}{\partial s} S_x(t,s)w = -S_x(t,s)A(s, x(s))w$, $w \in Y$.

2.1. Motivation

2.1.1. Theory of Elastic String

Consider the following integro-differential equation

$$\frac{\partial^2 z(x,t)}{\partial t^2} + c(t)\frac{\partial z(x,t)}{\partial t} - M\left(\int_{-\infty}^{+\infty} |\frac{\partial z(x,s)}{\partial s}|^2 ds\right)\frac{\partial^2 z(x,t)}{\partial x^2} + z(t,x)$$
$$= l(t, x, z(t,x)), \ 0 \leq t < \infty, \ x \in \mathcal{R},$$
$$z(0,x) = z_0(x),$$
$$\frac{\partial z(x,0)}{\partial t} = z_1(x), \ x \in \mathcal{R}.$$

Equations of this type occur in the study of the nonlinear behavior of an elastic string [15]. The basic physical assumptions are that the longitudinal strain of the string is very small and that the tension F is uniform along the string but may vary with time to accommodate changes in the arc length of the string. The nonlinearity arises from the assumption that F depends on the arc length S of the string at time $t \leq 0$ by the relation $F = F_0 + C[(S-L)/L]$ where F_0 is the minimum tension, L is the minimum length and C is a physical constant.

There are other types of integro-differential equations, similar to the above, which occur in the study of dynamical buckling of the hinged extensible beam which is either stretched or compressed by axial force. An equation of the following form

$$u_t(t,x) + \Psi(u(t,x))_x = \int_0^t b(t-s)\Psi(u(s,x))_x ds + f(t,x), \ t \in [0,a], x \in \mathcal{R},$$
$$u(0,x) = \phi(x), \ x \in \mathcal{R},$$

occurs in a nonlinear conversion law with memory.

The above type of equation can be formulated abstractly as

$$\dot{x}(t) + A(t, x(t))x(t) = Bu(t) + f(t, x(t), \int_0^t h(t,s,x(s))ds), \ t \in [0,a],$$
$$x(0) = x_0,$$

where $-A$ is the infinitesimal generator of an analytic semigroup of linear operators and f, h are given nonlinear functions.

2.1.2. Diffusion and Heat Conduction

In the theory of diffusion and heat conduction, one encounters a mathematical model of the form [16]

$$Lu + c(x,t)u = f(x,t), \ x \in \Omega, \ 0 < t < a,$$
$$u(x,t) = \phi(x,t), \ x \in \partial\Omega, \ 0 < t < a,$$
$$u(x,0) + \sum_{k=1}^{n} \beta_k(x, a_k) = \psi(x), \ x \in \Omega,$$

with $a_k \in (0, a]$ $(k = 1, 2 \cdots, N)$, where Ω is a bounded domain in $\mathcal{R}^n$ and L is a uniformly parabolic operator with continuous and bounded coefficients. It represents the diffusion phenomenon of a small amount of gas in a transparent tube. If there is very little gas at the initial time, the measurement $u(x, 0)$ of the amount of the gas in this instant may be less precise than the measurement $u(x,0) + \sum_{k=1}^{n} \beta_k(x, a_k)$ of the amount of this gas.

2.1.3. Quasi-Linear Differential Equations

Abstract quasi-linear integro-differential equations arise in many areas of science such as population dynamics, mathematical physics, heat conduction theory of material with memory etc. For this reason, this type of equations has received much attention in recent years. The literature related to quasi-linear differential and integro-differential equations is very extensive.

The Cauchy problem for the quasi-linear initial value problem in a Banach space X of the form

$$\frac{du(t)}{dt} + A(t, u)u = 0, \ \text{ for } 0 \le t \le a,$$
$$u(0) = u_0,$$

where the linear operator $A(t, u)$ appearing in the problem depends explicitly on the solution u of the problem, while in the semi-linear case the nonlinear operator is the sum of a fixed linear operator (independent of the solution u) and a nonlinear "function" of u.

In general, the study of quasi-linear initial value problems is quite complicated. We begin by briefly indicating the general idea behind the definition and the existence proof of a mild solution.

Let $u \in \mathcal{C}([0, a] : X)$ and consider the linear initial value problem

$$\frac{dv(t)}{dt} + A(t, u)v = 0, \quad \text{for } 0 \le t \le a,$$

$$v(0) = u_0.$$

If this problem has a unique mild solution $v \in \mathcal{C}([0, a] : X)$, for every given $u \in \mathcal{C}([0, a] : X)$, then it defines a mapping $u \to v = F(u)$ of $\mathcal{C}([0, a] : X)$ into itself. The fixed point of this mapping is defined to be a mild solution of (2.1, 2.2).

To prove the existence of a mild solution of (2.1, 2.2), we will show that, under suitable conditions, there always exists a T_0, $0 < T_0 \le a$, such that the restriction of the mapping F to $\mathcal{C}([0, T_0] : X)$ is a contraction which maps some ball of $\mathcal{C}([0, T_0] : X)$ into itself. The contraction mapping principle will then imply the existence of a unique fixed point u of F in this ball and u is then, by definition, the desired mild solution of (2.1, 2.2).

2.2. Methods

2.2.1. Semigroup Theory

The theory of semigroups of bounded linear operators is closely related to the solution of differential and integro-differential equations in Banach spaces. This theory has developed quite rapidly since the discovery of the generation theorem by Hille and Yosida in 1948. By now, it is an extensively studied mathematical subject with substantial applications to many fields of analysis. Pazy [1] discussed the existence and uniqueness of evolution equations using semigroup theory and fixed point theorems. By using the abstract approach, it is possible to extend the analysis developed on finite-dimensional linear systems to infinite-dimensional linear systems. In recent years, the method of semigroups has been applied to study the controllability problems for a large class of nonlinear differential and integro-differential evolution systems in Banach spaces. Many problems in the fields of ordinary and partial differential equations can be recast as integral equations. Several existence and uniqueness results can be derived from the corresponding results of integral equations. Such results can be obtained by applying the fixed-point theorems.

2.2.2. Fixed Point Method

Of all the methods, the fixed point method is the most effective one to study the controllability of nonlinear integro-differential systems. The fixed point method is used to prove the existence of theorems for integro-differential equations and study the controllability problem for integro-differential systems. In this method, the problem is transformed into a fixed point problem for an appropriate nonlinear operator in a function space. Moreover, by using the fixed point theorems, one can obtain controllability conditions in the Banach spaces of continuous functions. We mainly employ the Sadovskii's fixed point theorem, Banach contraction principle, and Schauder's fixed point theorem to study the controllability results for nonlinear neutral impulsive integro-differential evolution systems in Banach spaces.

2.2.3. Measure of Noncompactness

Measures of non-compactness play an important role in nonlinear analysis. They are often applied to the theories of differential and integral equations as well as to the operator theory and geometry of Banach spaces. The concept of a measure of non-compactness was initiated by Kuratowski and Darbo. Measures of non-compactness form a handy tool in many branches of mathematics. There are two important measures in which Kuratowski introduced the first measure of non-compactness $\alpha(X)$ of a bounded set X in a metric space defined as the infimum of $d > 0$ such that X can be covered with a finite number of sets of diameters smaller than d. Another important and very convenient measure is the so-called Hausdorff (or ball of measure) measure of non-compactness $\mu(X)$ defined by

$$\mu(X) = \inf\{d > 0 : X \text{ can be covered by a finite number of balls of radii smaller than } d\}.$$

The Hausdorff measure is frequently used in many branches of nonlinear analysis and its applications. The fact that it is defined naturally and has several very useful properties. Roughly speaking, the measure of non-compactness is in almost all these definitions. Some functions are defined on the family of all bounded and nonempty subsets of a given metric space such that it is equal to zero on the whole family of relatively compact sets. The compactness conditions described by means of measures of non-compactness are useful in the controllability results for differential and integro-differential systems in Banach spaces.

In the sequel, this work in the space $C(J, X)$ consisting of all functions defined and continuous on J with values in Banach space X. The space $C(J, X)$ is furnished with standard norm $\|x\|_C = \sup\{\|x(t)\| : t \in J\}$. In order to define the measure let us fix a nonempty bounded subset Y of the space $C(J, X)$ and a positive number $t \in J$. For $y \in Y$ and $\epsilon \geq 0$ denoted by $\omega^t(y, \epsilon)$ the *modulus of continuity* of the function y on the interval $[0, t]$, that is

$$\omega^t(y, \epsilon) = \sup\{\|y(t_2) - y(t_1)\| : t_1, t_2 \in [0, t], |t_2 - t_1| \leq \epsilon\}.$$

Further, let us put

$$\omega^t(Y, \epsilon) = \sup\{\omega^t(y, \epsilon) : y \in Y\},$$
$$\omega_0^t(Y) = \lim_{\epsilon \to 0+} \omega^t(Y, \epsilon).$$

Apart from this, put $\overline{\beta}(Y) = \sup\{\beta(Y(t)) : t \in J\}$, where β denotes Hausdorff measure of non-compactness in X. Finally, we define a function χ on the family $\mathcal{F}_{C(J,X)}$ by putting

$$\chi(Y) = \omega_0^t(Y) + \overline{\beta}(Y).$$

It may be shown that the function χ is the measure of non-compactness in space $C(J, X)$. The kernel kerχ is the family of all nonempty and bounded subsets Y such that functions belonging to Y are equi-continuous on J and the set $Y(t)$ is relatively compact in X, $t \in J$.

Next, for a given set $Y \in \mathcal{F}_{C(J,X)}$, let us denote

$$\int_0^t Y(s)ds = \left\{\int_0^t y(s)ds : y \in Y\right\}, \quad t \in J,$$
$$Y([0, t]) = \{y(s) : y \in Y, s \in [0, t]\}.$$

Lemma 2.1. *If the Banach space X is separable and a set $Y \subset C(J, X)$ is bounded, then the function $t \to \beta(Y(t))$ is measurable and*

$$\beta\left(\int_0^t Y(s)ds\right) \leq \int_0^t \beta(Y(s))ds, \quad \text{for each } t \in J.$$

Lemma 2.2. *Assume that a set* $Y \subset C(J, X)$ *is bounded. Then*

$$\beta(Y([0,t])) \leq \omega_0^t(Y) + \sup_{s \leq t} \beta(Y(s)), \text{ for } t \in J. \tag{2.4}$$

Proof. For arbitrary $\delta > 0$. Then there exists $\epsilon > 0$ such that

$$\omega^t(Y, \epsilon) \leq \omega_0^t(Y) + \delta/2. \tag{2.5}$$

Let us take a partition $0 = t_0 < t_1 < ... < t_k = t$ such that $t_i - t_{i-1} \leq \epsilon$, for $i = 1, 2, ..., k$. Then for each $t' \in [t_{i-1}, t_i]$ and $y \in Y$ the following inequality is fulfilled

$$\|y(t') - y(t_i)\| \leq \omega_0^t(Y) + \delta/2. \tag{2.6}$$

Let us notice that, for each $i = 1, 2, ..., k$ there are points $z_{ij} \in X$ ($j = 1, 2, ..., n_i$) such that

$$Y(t_i) \subset \bigcup_{j=1}^{n_i} B(z_{ij}, \sup_{s \leq t} \beta(Y(s)) + \delta/2). \tag{2.7}$$

We show that

$$Y([0,t]) = \bigcup_{i=1}^{k} \bigcup_{j=1}^{n_i} B(z_{ij}, \sup_{s \leq t} \beta(Y(s)) + \omega_0^t(Y) + \delta/2). \tag{2.8}$$

Let us choose an arbitrary element $q \in Y([0,t])$. Then, we can find $t' \in [0,t]$ and $y \in Y$, such that $q = y(t')$. Choosing i such that $t' \in [t_{i-1}, t_i]$ and j such that $B(z_{ij}, \sup_{s \leq t} \beta(Y(s)) + \delta/2)$, we obtain from (2.5) and (2.6)

$$\|q - z_{ij}\| = \|y(t') - z_{ij}\| \leq \|y(t') - y(t_i)\| + \|y(t_i) - z_{ij}\|$$
$$\leq \omega_0^t(Y) + \sup_{s \leq t} \beta(Y(s)) + \delta$$

and this verifies (2.7). Condition (2.7) yields that

$$\beta(Y([0,t])) \leq \omega_0^t(Y) + \sup_{s \leq t} \beta(Y(s)) + \delta.$$

Letting $\delta \to 0+$ we get (2.4). $\square$

Definition 2.3. *A solution $x(\cdot) \in C([0,b], X)$ is said to be a mild solution of (2.1, 2.2) if $x(s) = g(x), s \in [0,b]$, then for each $0 \leq s \leq t \leq b$, and the following integral equation is satisfied.*

$$x(t) = S_x(t,0)g(x) + \int_0^t S_x(t,s)Bu(s)ds + \int_0^t S_x(t,s)f(s,x(s))ds, \ t \in J.$$

To study the controllability problem, we assume the following hypotheses:

H1. $A(t,s)$ generates strongly continuous semigroup of a family of evolution operators $S_x(t,s)$ and there exist constants $N_1 > 0$, $N_0 > 0$ such that

$$\|S_x(t,s)\| \leq N_1, \quad \text{for } 0 \leq s \leq t \leq b,$$

and

$$N_0 = \sup\{\|S_x(s,0)\| : 0 \leq s \leq t\}.$$

H2. The linear operator $W : L^2(J,U) \to X$ defined by

$$Wu = \int_0^b S_x(b,s)Bu(s)ds$$

has an inverse operator W^{-1}, which takes values in $L^2(J,U)/\ker W$ and there exists a positive constant K_1 such that $\|BW^{-1}\| \leq K_1$.

H3. (i) The mapping $f : J \times X \times X \to X$ satisfies the Caratheodory condition, a.e., $f(\cdot, x)$ is measurable for $x \in X$ and $f(t, \cdot)$ is continuous, for $t \in J$.

(ii) The mapping f is bounded on bounded subsets of $C(J,X)$.

(iii) There exists a constant $m_f > 0$ such that for any bounded set $Y \subset C(J,X)$, the inequality

$$\beta(f([0,t] \times Y)) \leq m_f \beta(Y([0,t]))$$

holds for $t \in J$, where

$$f([0,t] \times Y) = \{f(s,x(s)) : 0 \leq s \leq t, \ x \in Y\}.$$

H4. The function $g : C(J,X) \to X$ is continuous and there exists a constant $m_g \geq 0$ such that

$$\beta(g(Y)) \leq m_g \beta(Y(J)),$$

for each bounded set $Y \subset C(J,X)$.

H5. There exists a constant $r > 0$ such that

$$(1+bN_1K_1)\left[N_0 \sup_{x\in\mathbb{B}(\theta,r)} \|g(x)\| + N_1 \sup_{x\in\mathbb{B}(\theta,r)} \int_0^b \|f(\tau,x(\tau))\|d\tau\right] + bN_1K_1\|x_1\| \leq r,$$

for $t \in J$, where $\mathbb{B}(\theta, r)$ is closed ball in $\mathcal{C}(J, X)$ centered at θ and with radius r.

Definition 2.4. *[8, 17, 18] System (2.1, 2.2) is said to be controllable on the interval J, if for every initial function $x_0 \in X$ and $x_1 \in X$, there exists a control $u \in L^2(J,U)$ such that the solution $x(\cdot)$ of (2.1, 2.2) satisfies $x(0) = x_0$ and $x(b) = x_1$.*

3. CONTROLLABILITY RESULT

Mathematical control theory is the area of application-oriented mathematics that deals with the basic principles underlying the analysis and design of control systems. Controllability is an important property of a control system and the controllability property plays a crucial role in many control problems such as stabilization of unstable systems with feedback or optimal control. To control an object means to influence its behavior to achieve the desired goal. Apart from these, the study of controllability and observability properties of a system in control theory is certainly, at present, one of the most active interdisciplinary areas of research. On the other hand, control theory has remained a discipline where many mathematical ideas and methods have fused to produce a new body of important mathematics.

In this section, we study the controllability results for the quasi-linear differential system (2.1, 2.2). Using H2 for an arbitrary function $x(\cdot) \in \mathcal{C}(J, X)$, we define the control

$$u(t) = W^{-1}\left[x_1 - S_x(b,0)g(x) - \int_0^b S_x(b,s)f(s,x(s))ds\right](t). \quad (3.1)$$

Consider the Banach space $\mathcal{Z} = \mathcal{C}(J, X)$ with norm $\|x\| = \sup\{|x(t)| : t \in J\}$. We shall show that when using the control $u(t)$, the operator $\Psi : \mathcal{Z} \to \mathcal{Z}$ defined by

$$(\Psi x)(t) = S_x(t,0)g(x) + \int_0^t S_x(t,s)f(s,x(s))ds$$

$$+ \int_0^t S_x(t,s)BW^{-1}\left[x_1 - S_x(b,0)g(x) - \int_0^b S_x(b,\tau)f(\tau,x(\tau))d\tau\right](s)ds$$

has a fixed point $x(\cdot)$. This fixed point is the mild solution of the system (2.1, 2.2), which implies that the system is controllable on J.

Next, consider the operators $v_1, v_2, v_3 : C(J, X) \to C(J, X)$ defined by

$$(v_1 x)(t) = S_x(t, 0)g(x)$$

$$(v_2 x)(t) = \int_0^t S_x(t, s)f(s, x(s))ds$$

$$(v_3 x)(t) = \int_0^t S_x(t, s)BW^{-1}\left[x_1 - S_x(b, 0)g(x) - \int_0^b S_x(b, \tau)f(\tau, x(\tau))d\tau\right](s)ds.$$

Lemma 3.1. *Assume that assumptions H1, H4 are satisfied and a set $Y \subset C(J, X)$ is bounded. Then*

$$\omega_0^t(v_1 Y) \leq 2N_0(t)\beta(g(Y)), \ \text{for} \ t \in J.$$

The simple proof is omitted.

Lemma 3.2. *Assume that H1 and H3 are satisfied and a set $Y \subset C(J, X)$ is bounded. Then*

$$\omega_0^t(v_2 Y) \leq 2bN_1\beta(f([0, b] \times Y)), \ \text{for} \ t \in J.$$

Proof. Fix $t \in J$ and denote $Q = f([0, t] \times Y)$,

$$q^t(\epsilon) = \sup\left\{\|(S_x(t_2, s) - S_x(t_1, s))q\| : 0 \leq s \leq t_1 \leq t_2 \leq t, \ t_2 - t_1 \leq \epsilon, \ q \in Q\right\}.$$

At the beginning, we show that

$$\lim_{\epsilon \to 0+} q^t(\epsilon) \leq 2N_1\beta(Q). \tag{3.2}$$

Suppose contrary, then there exists a number d such that

$$\lim_{\epsilon \to 0+} q^t(\epsilon) > d > 2N_1\beta(Q). \tag{3.3}$$

Fix $\delta > 0$ such that

$$\lim_{\epsilon \to 0+} q^t(\epsilon) > d + \delta > d > 2N_1(\beta(Q) + \delta). \tag{3.4}$$

Condition (3.2) yields that there exist sequences $\{t_{2,n}\}, \{t_{1,n}\}, \{s_n\} \subset J$ and $\{q_n\} \subset Q$, such that $t_{2,n} \to t'$, $t_{1,n} \to t'$, $s_n \to s$ and

$$\|(S_x(t_{2,n}, s_n) - S_x(t_{1,n}, s_n))q_n\| > d. \tag{3.5}$$

Let the points $z_1, z_2, ..., z_k \in X$ be such that $Q \subset \bigcup_{i=1}^{k} B(z_i, \beta(Q) + \delta)$. Then there exists a point z_i and a sub-sequence $\{q_n\}$, such that $\{q_n\} \in B(z_i, \beta(Q) + \delta)$, that is,
$$\|z_j - q_n\| \leq \beta(Q) + \delta, \text{ for } n = 1, 2, ...$$

Further, we obtain
$$\|S_x(t_{2,n}, s_n)q_n - S_x(t_{1,n}, s_n)q_n\| \leq 2N_1(\beta(Q) + \delta) + \|S_x(t_{2,n}, s_n)z_j - S_x(t_{1,n}, s_n)z_j\|.$$

Letting $n \to \infty$, and using the properties of the evolution system $\{S_x(t,s)\}$, from the above estimation, we get
$$\limsup_{n\to\infty} \|S_x(t_{2,n}, s_n)q_n - S_x(t_{1,n}, s_n)q_n\| \leq 2N_1(\beta(Q) + \delta).$$

This contradicts (3.1) and (3.2).

Now, fix $\epsilon > 0$ and $t_1, t_2 \in [0, t]$, $0 \leq t_2 - t_1 \leq \epsilon$. Applying H3, we get

$$\|(v_2 x)(t_2) - (v_2 x)(t_1)\| \leq \int_0^{t_1} \|(S_x(t_2, s) - S_x(t_1, s))f(s, x(s))\|ds$$
$$+ \int_{t_1}^{t_2} \|S_x(t_2, s))f(s, x(s))\|ds$$
$$+ \int_0^{t_2} \|(S_x(t_2, s) - S_x(t_1, s))f(s, x(s))\|ds$$
$$+ \epsilon N_1 \sup\{\|f(s, x(s))\| : x \in Y\}.$$

Hence, we derive the following inequality

$\omega^t(v_2 Y, \epsilon)$
$$\leq \sup\left\{\int_0^t \|(S_x(t_2, s) - S_x(t_1, s))f(s, x(s))\|ds : t_1, t_2 \in [0, t], 0 \leq t_2 - t_1 \leq \epsilon, x \in Y\right\}$$
$$+ \epsilon N_1 \sup\{\|f(s, x(s))\| : x \in Y\}.$$

Letting $\epsilon \to 0+$ and hence the result. $\square$

Lemma 3.3. *Assume that assumptions H1-H4 are satisfied and a set $Y \subset C(J, X)$ is bounded. Then*
$$\omega_0^t(v_3 Y) \leq 2bN_1 K_1\Big(\|x_1\| + N_0\beta(g(Y)) + bN_1\beta(f(Q))\Big), \text{ for } t \in J.$$

Proof. From the above Lemmas 3.1 and 3.2, also fix $\epsilon > 0$ and $t_1, t_2 \in [0,t]$, $0 \leq t_2 - t_1 \leq \epsilon$. Applying H3 and H4, we get

$$\|(v_3x)(t_2) - (v_3x)(t_1)\|$$
$$\leq K_1 \int_0^{t_1} \|S_x(t_2,s) - S_x(t_1,s)\| \Big[\|x_1\| + \|S_x(b,0)g(x)\| + \int_0^b \|S_x(b,\tau)f(\tau,x(\tau))d\tau\|\Big] ds$$
$$+ \epsilon K_1 N_1 \Big[\|x_1\| + N_0 \sup\{\|g(x)\| : x \in Y\} + N_1 \sup\{\|f(s,x(s))\| : x \in Y\}\Big].$$

Hence we derive the following inequality

$$\omega_0^t(v_3 Y) \leq \sup \Big\{ K_1 \int_0^{t_1} \|S_x(t_2,s) - S_x(t_1,s)\| \Big[\|x_1\| + \|S_x(b,0)g(x)\|$$
$$+ \int_0^b \|S_x(b,\tau)f(\tau,x(\tau))d\tau\|\Big] ds : t_1, t_2 \in [0,b],\ 0 \leq t_2 - t_1 \leq \epsilon, x \in Y \Big\}$$
$$+ \epsilon K_1 N_1 \Big[\|x_1\| + N_0 \sup\{\|g(x)\| : x \in Y\} + N_1 \sup\{\|f(s,x(s))\| : x \in Y\}\Big].$$

Letting $\epsilon \to 0+$, we get

$$\omega_0^t(v_3 Y) \leq 2bN_1 K_1 \Big(\|x_1\| + N_0 \beta(g(Y)) + bN_1 \beta(f(Q))\Big). \quad \square$$

Then we can calculate our main result as follows:

Theorem 3.1. *If the Banach space X is separable under the assumptions H1-H4, the system (2.1, 2.2) is controllable on J.*

Proof. Consider the operator $\mathcal{P}$ defined by

$$(\mathcal{P}x)(t) = S_x(t,0)g(x) + \int_0^t S_x(t,s)f(s,x(s))ds$$
$$+ \int_0^t S_x(t,s)BW^{-1}\Big[x_1 - S_x(b,0)g(x) - \int_0^b S_x(b,s)f(s,x(\tau))d\tau\Big](s)ds.$$

For an arbitrary $x \in \mathcal{C}(J,X)$ and $t \in J$, we get

$$\|(\mathcal{P}x)(t)\| \leq (1 + bN_1 K_1)\Big[N_0\|g(x)\| + N_1 \int_0^b \|f(\tau,x(\tau))\|d\tau\Big] + bN_1 K_1 \|x_1\| \leq r.$$

Now, we prove the operator $\mathcal{P}$ is continuous on $\mathbb{B}(\theta,r)$.

Let us fix $x \in \mathbb{B}(\theta,r)$ and take arbitrary sequence $\{x_n\} \in \mathbb{B}(\theta,r)$ such that $x_n \to x$ in $\mathcal{C}(J,X)$. Next, we have

$$\|\mathcal{P}x_n - \mathcal{P}x\| \leq (1 + bN_1 K_1)\Big[N_0\|g(x_n) - g(x)\| + N_1 \int_0^b \|f(\tau,x_n(\tau)) - f(\tau,x(\tau))\|d\tau\Big].$$

Applying Lebesgue dominated convergence theorem, we derive that $\mathcal{P}$ is continuous on $\mathbb{B}(\theta, r)$.

Now, we consider the sequence of sets $\{\Omega_n\}$ defined by induction as follows:
$$\Omega_0 = \mathbb{B}(\theta, r), \ \Omega_{n+1} = Conv(\mathcal{P}\Omega_n), \text{ for } n = 1, 2, \ldots.$$

This sequence is decreasing, i.e., $\Omega_n \supset \Omega_{n+1}$, for $n = 1, 2, \ldots$.

Further let us put $v_n(t) = \beta(\Omega_n([0, t]))$ and $w_n(t) = \omega_0^t(\Omega_n)$. Observe that each of functions $v_n(t)$ and $w_n(t)$ is nondecreasing, while sequences $\{v_n(t)\}$ and $\{w_n(t)\}$ are nonincreasing at any fixed $t \in J$. Put

$$v_\infty(t) = \lim_{n \to \infty} v_n(t),$$
$$w_\infty(t) = \lim_{n \to \infty} w_n(t), \text{ for } t \in J.$$

Using Lemmas 2.2, 3.2 and H4, we obtain

$$\beta(v_1\Omega_n([0, t])) \leq \omega_0^t(v_1\Omega_n) + \sup_{s \leq t} \beta(v_1\Omega_n(s)) = 3m_g N_0(t) v_n(b),$$

that is,

$$\beta(v_1\Omega_n([0, t])) \leq 3m_g N_0(t) v_n(b). \tag{3.6}$$

Moreover,

$$\beta(v_2\Omega_n([0, t])) \leq \omega_0^t(v_2\Omega_n) + \sup_{s \leq t} \beta(v_2\Omega_n(s))$$
$$\leq 2m_f b N_1(t) v_n(t) + m_f N_1(t) \int_0^t v_n(\tau) d\tau$$

and

$$\beta(v_3\Omega_n([0, t]))$$
$$\leq \omega_0^t(v_3\Omega_n) + \sup_{s \leq t} \beta(v_3\Omega_n(s))$$
$$\leq 3bN_1(t)K_1\Big(\|x_1\| + m_g N_0(t) v_n(b)\Big) + bm_f N_1(t) K_1\Big(2bN_1(t)v_n(t) + N_1 \int_0^t v_n(\tau) d\tau\Big)$$

Linking this estimate with (3.5), we obtain

$$v_{n+1}(t) = \beta(\Omega_{n+1}([0,t])) = \beta(\mathcal{P}\Omega_n([0,t]))$$
$$\leq 3m_g N_0(t) v_n(b) + 2m_f b N_1(t) v_n(t) + m_f N_1(t) \int_0^t v_n(\tau) d\tau$$
$$+ 3b N_1(t) K_1 \Big(\|x_1\| + m_g N_0(t) v_n(b) \Big)$$
$$+ b m_f N_1(t) K_1 \Big(2b N_1(t) v_n(t) + N_1 \int_0^t v_n(\tau) d\tau \Big).$$

Letting $n \to \infty$, we get

$$v_\infty(t) \leq 3m_g N_0(t) v_\infty(b) + 2m_f b N_1(t) v_\infty(t) + m_f N_1(t) \int_0^t v_\infty(\tau) d\tau$$
$$+ 3b N_1(t) K_1 \Big(\|x_1\| + m_g N_0(t) v_\infty(b) \Big)$$
$$+ b m_f N_1(t) K_1 \Big(2b N_1(t) v_\infty(t) + N_1 \int_0^t v_\infty(\tau) d\tau \Big).$$

Hence, puttting $t = b$, we get $v_\infty(b) = 0$. Furthermore, applying Lemmas 3.1, 3.2, 3.3, we have

$$w_{n+1}(t) = \omega_0^t(\Omega_{n+1}) = \omega_0^t(\mathcal{P}\Omega_n)$$
$$\leq (2 + b N_1 K_1)[m_g N_0 v_n(b) + m_f b N_1 v_n(t)] + 2b N_1 K_1 \|x_1\|.$$

Letting $n \to \infty$, we get

$$w_\infty(t) \leq (2 + b N_1 K_1)[m_g N_0 v_\infty(b) + m_f b N_1 v_\infty(t)] + 2b N_1 K_1 \|x_1\|.$$

Putting $t = b$ and applying (3.6), we conclude that $w_\infty(b) = 0$. This fact together with (3.6) implies that $\lim_{n \to \infty} \chi(\Omega_n) = 0$. Hence, in view of Remark, we deduce that the set $\Omega_\infty = \bigcap_{n=0}^\infty \Omega_n$ is nonempty, compact, and convex. Finally linking above obtained facts concerning the set Ω_∞ and the operator $\mathcal{P} : \Omega_\infty \to \Omega_\infty$ and using the classical Schauder fixed point theorem, we infer that the operator $\mathcal{P}$ has at least one fixed point x in the set Ω_∞. Obviously the function $x = x(t)$ is a mild solution of (2.1, 2.2) satisfying $x(b) = x_1$. Hence the given system is controllable on J. □

REFERENCES

[1] Pazy A. (1983), *Semigroups of Linear Operators and Applications to Partial Differential Equations*. Springer-Verlag, New York.

[2] Byszewski L. (1991), Theorems about the existence and uniqueness of solutions of a semi-linear evolution non-local Cauchy problem. *Journal of Mathematical Analysis and Applications*, 162, 494-505.

[3] Lin Y., Liu J. H. (2003), Semilinear integro-differential equations with non-local Cauchy problem. *Journal of Integral Equations and Applications*, 15, 79-93.

[4] Xue X. (2009), Non-local nonlinear differential equations with measure of non-compactness in Banach spaces. *Nonlinear Analysis: Theory Methods & Applications*, 70, 2593-2601.

[5] Klamka J. (1993), *Controllability of Dynamical Systems*. Kluwer Academic, Dordrecht.

[6] Chukwu E. N., Lenhart S. M. (1991), Controllability questions for nonlinear systems in abstract spaces. *Journal of Optimization Theory and Applications*, 68, 437-462.

[7] Curtain R. F., Zwart H. (1995), *An Introduction to Infinite Dimensional Linear Systems Theory*. Springer-Verlag, Berlin.

[8] Balachandran K., Dauer J. P. (1987), Controllability of nonlinear systems via Fixed Point Theorems. *Journal of Optimization Theory and Applications*, 53, 345-352.

[9] Chang Y. K., Nieto J. J., Li W. S. (2009), Controllability of semi-linear differential systems with nonlocal initial conditions in Banach spaces. *Journal of Optimization Theory and Applications*, 142, 267-273.

[10] Yan Z. (2007), Controllability of semilinear integro-differential systems with nonlocal conditions. *International Journal of Computational and Applied Mathematics*, 3, 221-236.

[11] Balachandran K., Dauer J. P. (2002), Controllability of nonlinear systems in Banach spaces: A survey, *Journal of Optimization Theory and Applications*, 115, 7-28.

[12] Naito K. (1992), On controllability for a nonlinear Volterra equation. *Nonlinear Analysis: Theory, Methods and Applications*, 18, 99-108.

[13] Naito K. (1987), Controllability of semilinear control systems dominated by the linear part. *SIAM Journal on Control and Optimization*, 25, 715-722.

[14] Banas J., Goebel K. (1980), *Measure of Non-compactness in Banach Space, in: Lecture Notes in Pure and Applied Mathematics*. Dekker, New York.

[15] Heard M. L. (1984), A quasilinear hyperbolic integro-differential equation related to a nonlinear string. *Transactions of American Mathematical Society*, 285, 805-823.

[16] Deng K. (1993), Exponential decay of solutions of semilinear parabolic equations with non-local initial conditions. *Journal of Mathematical Analysis and Applications*, 179, 630-637.

[17] Radhakrishnan B. (2015), Controllability Results for Nonlinear Neutral Fuzzy Integrodifferential Systems in Fuzzy Semigroups. *Journal of Nonlinear Functional Analysis*, 2015, 1-19.

[18] Radhakrishnan B., Balachandran K. (2012), Controllability results for semilinear impulsive integro-differential evolution systems with non-local conditions. *Journal of Control Theory and Applications*, 10, 28-34.

In: Hilbert Spaces: Properties and Applications
Editor: Le Bin Ho
ISBN: 978-1-53616-633-0
© 2020 Nova Science Publishers, Inc.

Chapter 4

DERIVATIONS OF OPERATOR ALGEBRAS ON HYPERCOMPLEX HILBERT SPACES AND RELATED MODULES

*S. V. Ludkowski**
Department of Applied Mathematics,
Moscow State Technological University MIREA-RTU,
Moscow, Russia

Abstract

Derivations of operator algebras and C^*-algebras over hypercomplex numbers are studied. Related modules over octonions and Cayley-Dickson algebras are investigated. Continuity of derivations, inner and outer derivations are scrutinized. Moreover cohomologies of such algebras and their generalizations are studied.

Keywords: Hilbert space, hypercomplex numbers, octonion algebra, operator, C^*-algebra, derivation, automorphism, Cayley-Dickson algebra, von Neumann algebra, infinite dimension

1. INTRODUCTION

Hilbert spaces over the real and complex fields play a significant role in mathematics and its applications. On the other hand, their hypercomplex generaliza-

---
*Corresponding Author's E-mail: sludkowski@mail.ru.

tions become important in recent years. Among them, nonassociative Cayley-Dickson algebras are intensively studied. They are generalizations of the octonion algebra (see [1–5] and references therein). They are formed by induction using a doubling procedure of a smashed product. Their structure and identities in them attract great attention (see, for example, [2, 4, 6–9] and references therein). They have found many-sided applications in the theory of Lie groups and algebras and their generalizations, mathematical analysis, non-commutative geometry, operator theory, PDE and their applications in natural sciences including physics and quantum field theory (see [7, 10–28] and references therein).

Derivations of finite dimensional Cayley-Dickson algebras, particularly, of the octonion algebra also, were described in [5, 19, 29, 30]. On the other hand, derivations of infinite dimensional associative operator algebras were investigated.

It is worth to mention that for algebras A over rings or fields K derivations d are supposed to be linear over rings or fields:

$$d(a + b) = da + db, \tag{1.1}$$

$$d(ab) = (da)b + adb, \text{ and} \tag{1.2}$$

$$d(ta) = tda, \tag{1.3}$$

for each a and b in A and t in K. This definition for unital algebras implies that $dt = 0$ for each $t \in K$ since $d1 = d(11) = 2d1$, where K is embedded into A as $K1$. Such a convention may be apart from derivations of groups or rings (or fields in particular) G considered in some other books and articles, where instead of three only two conditions (1.1), (1.2) are imposed for each a and b in G and with values da in a chosen module (see, for example, [31]).

Derivations of algebras appear as particular cases of cohomologies of algebras. Cohomology theory of associative algebras was investigated by Hochschild and other authors [32, 33], but it does not apply to nonassociative algebras studied in this chapter.

Moreover, in several works, it was elucidated that derivations of von Neumann algebras and simple C^*-algebras on complex Hilbert spaces are inner [34, 35]. The latter algebras are associative. They have specific features in comparison with the case of complex manifolds. Derivations of infinite dimensional associative algebras using their cohomologies were studied in [36]. But this is not valid in the nonassociative case as it is shown in this chapter. Note that derivations and cohomologies of infinite dimensional associative algebras have

Derivations of Operator Algebras on Hypercomplex Hilbert Spaces ... 63

specific features in comparison with finite dimensional algebras. Moreover, cohomologies of modules over hypercomplex algebras have specific features in comparison with the case of modules over complex algebras. This is especially caused by the noncommutativity of the quaternion skew field and the nonassociativity of the octonion algebra.

Previously, cohomologies of loop spaces on quaternion and octonion manifolds were studied in [21]. Spectral theory of operators over quaternions and octonions was presented in [22, 23, 25–27, 37].

In this chapter, derivations and automorphisms of operator algebras on Hilbert spaces over hypercomplex numbers are investigated. Particularly infinite dimensional Cayley-Dickson algebras are also considered. Related modules over octonions and Cayley-Dickson algebras are investigated. Different types of metrics and norms on them are studied. It is proved below that there exist nonassociative analogs of von Neumann and simple C^*-algebras which have outer derivations and automorphisms. The corresponding cohomology theory is developed for this purpose. It is studied below even for more general metagroup algebras. Moreover, homomorphisms, continuity of derivations, inner and outer derivations and automorphisms of infinite dimensional Cayley-Dickson algebras and of operator algebras over Cayley-Dickson algebras are investigated.

All the main results of this chapter are obtained for the first time. They elucidate structures of operator algebras in hypercomplex Hilbert spaces, spectra of operators, the structure of modules over hypercomplex algebras. Their applications are discussed.

2. Homomorphisms of Infinite Dimensional Hypercomplex Algebras

To avoid misunderstandings, we recall necessary definitions.

Definition 2.1. Suppose that F is an associative commutative and unital ring. Elements of F are frequently considered as scalars. Then over F a unital algebra A is considered, which may be generally nonassociative. Let A be supplied with a scalar involution $a \mapsto \bar{a}$ so that its norm N and trace T maps have values in

F and fulfill conditions:

$$a\bar{a} = N(a)1 \text{ with } N(a) \in F, \tag{2.1}$$
$$a + \bar{a} = T(a)1 \text{ with } T(a) \in F, \tag{2.2}$$
$$T(ab) = T(ba), \tag{2.3}$$

for each a and b in A.

If a scalar $f \in F$ satisfies the condition: $\forall a \in A, fa = 0 \Rightarrow a = 0$, then such element f is called cancellable. For a cancellable scalar f, the Cayley-Dickson doubling procedure provides new algebra $C(A, f)$ over F such that:

$$C(A, f) = A \oplus Al, \tag{2.4}$$
$$(a + bl)(c + dl) = (ac - f\bar{d}b) + (da + b\bar{c})l, \text{ and} \tag{2.5}$$
$$\overline{(a + bl)} = \bar{a} - bl, \tag{2.6}$$

for each a and b in A. Then l is called a doubling generator.

Remark 2.1. From Definition 2.1, there follows that $\forall a, b \in A$,

$$T(a) = T(a + bl), \text{ and}$$
$$N(a + bl) = N(a) + fN(b).$$

The algebra A is embedded into $C(A, f)$ as $A \ni a \mapsto (a, 0)$, where $(a, b) = a + bl$. Put by induction $A_n(f_{(n)}) = C(A_{n-1}, f_n)$, where $A_0 = A$, $f_1 = f$, $n = 1, 2, \cdots$, $f_{(n)} = (f_1, \cdots, f_n)$, then $A_n(f_{(n)})$ are generalized Cayley-Dickson algebras when F is not a field, or Cayley-Dickson algebras when F is a field.

Moreover, we define

$$A_\infty(f) := \bigcup_{n=1}^{\infty} A_n(f_{(n)}), \tag{2.7}$$

where $f = (f_n : n \in \mathbf{N})$.

When $\text{char}(F) \neq 2$, it is put by the definition

$$\text{Im}(z) = z - T(z)/2, \tag{2.8}$$

to be the imaginary part of a Cayley-Dickson number z and hence $N(a) := N_2(a, \bar{a})/2$, where $N_2(a, b) := T(a\bar{b})$.

If the doubling procedure starts from $A = F1 =: A_0$, then $A_1 = C(A, f_1)$ is a $*$-extension of F. If A_1 has a basis $\{1, u\}$ over F with the multiplication table $u^2 = u + w$, where $w \in F$ and $4w + 1 \neq 0$, with the involution $\bar{1} = 1$, $\bar{u} = 1 - u$, then A_2 is the generalized quaternion algebra, A_3 is the generalized octonion (Cayley-Dickson) algebra.

In particular, if $F = \mathbf{R}$ is the real field, then up to normalization of the doubling generator l_k on k-th step a scalar $f_k \in \{-1, 1\}$ can be chosen for each $k = 1, 2, \cdots$.

When $F = \mathbf{R}$ and $f_n = 1$ for each n then $\mathcal{A}_r$ will be denoted the real Cayley-Dickson algebra with generators $i_0, \cdots, i_{2^r-1}$ such that $i_0 = 1$, $i_j^2 = -1$ for each $j \geq 1$, $i_j i_k = -i_k i_j$ for each $j \neq k \geq 1$. Notice that, for example, $(i_1 i_2) i_4 = -i_1 (i_2 i_4)$.

The Cayley-Dickson algebra $\mathcal{A}_{r+1}$ is formed from the algebra $\mathcal{A}_r$ with the help of the doubling procedure by generator i_{2^r}. In particular, $\mathcal{A}_0 := \mathbf{R}$ is the real field, $\mathcal{A}_1 = \mathbf{C}$ coincides with the field of complex numbers, $\mathcal{A}_2 = \mathbf{H}$ is the skew field of quaternions, $\mathcal{A}_3$ is the algebra of octonions, $\mathcal{A}_4$ is the algebra of sedenions. The algebra $\mathcal{A}_r$ is power associative, that is,

$$z^{n+m} = z^n z^m, \tag{2.9}$$

for each $n, m \in \mathbf{N}$ and $z \in \mathcal{A}_r$. It is nonassociative and non-alternative for each $r \geq 4$.

The Cayley-Dickson algebras $\mathcal{A}_r$ are nicely normed, that is,

$$a + \bar{a} =: 2\mathrm{Re}(a) \in \mathbf{R}, \text{ and} \tag{2.10}$$

$$a\bar{a} = \bar{a}a > 0, \tag{2.11}$$

for each $0 \neq a \in \mathcal{A}_r$. The norm $|\cdot|$ in it is defined by the formula:

$$|a|^2 := a\bar{a} = N(a). \tag{2.12}$$

Frequently $\bar{a}$ is also denoted by a^* or $\tilde{a}$. Each non-zero Cayley-Dickson number $0 \neq a \in \mathcal{A}_r$ has the multiplicative inverse given by $a^{-1} = \bar{a}/|a|^2$. Evidently, $\mathrm{Re}(z) = T(z)/2$ in this particular case.

Theorem 2.1. Let F be a Banach associative commutative and unital ring with $\mathrm{char}(F) \neq 2$ and $A_0 = F$ and let the infinite product converge:

$$0 < \prod_{n, |f_n| \geq 1} |f_n| < \infty$$

1. Then for each $0 < p < \infty$ there exists an l_p completion $A_{\infty,p}(f)$ of $A_\infty(f)$ so that $A_{\infty,p}(f)$ is an algebra, which is metrizable.

2. Moreover, if a norm on F is non-archimedean such that $\Gamma_F := \{|b| : b \in F, b \neq 0\} \subset (0, \infty)$, then $A_\infty(f)$ has also the c_0 completion $A_{\infty,c_0}(f)$ so that it is an ultrametrizable algebra.

3. If additionally F is a field supplied with a multiplicative norm, then $A_{\infty,p}(f)$ is a Banach algebra in case 1 when $1 \le p < \infty$, or $A_{\infty,c_0}(f)$ is a Banach algebra in case 2.

Proof. Since $\text{char}(F) \neq 2$, the Cayley-Dickson algebra $A_n(f_{(n)})$ has a basis $\mathbf{b}_n := \{i_0, i_1, \cdots, i_{2^n-1}\}$ over F for each natural number n, where $i_0 = 1$. They are related by induction: a basis $\mathbf{b}_{n+1}$ of $A_{n+1}(f_{(n+1)})$ over F is obtained from the preceding $\mathbf{b}_n$ and a doubling generator i_{2^n} and products $i_k i_{2^n} =: i_{k+2^n}$ for each $k = 1, \cdots, 2^n - 1$.

For each element $z \in A_\infty(f)$ there are $z_j \in F$ for each $j = 0, 1, 2, \cdots$ so that $z = \sum_j z_j i_j$, where only a finite number of z_j are different from zero.

1. We put

$$|z|_p := \left(\sum_j |z_j|^p\right)^{1/p}, \qquad (2.13)$$

where $0 < p < \infty$. In view of Section 4.2 in [38] the inequalities are valid

$$|y+z|_p \le |y|_p + |z|_p, \qquad (2.14)$$

for each $1 \le p$, also

$$|y+z|_p^p \le |y|_p^p + |z|_p^p, \qquad (2.15)$$

for each $0 < p < \infty$ and for all y and z in $A_\infty(f)$ since $|z_j| \ge 0$ for each j. Evidently, $|z|_p = 0$ if and only if $z = 0$; also $|tz|_p \le |t||z|_p$ for every $t \in F$ and $z \in A_\infty(f)$ since $|tz_j| \le |t||z_j|$ for each j. That is $A_\infty(f)$ is a normed two-sided F-module when $1 \le p$ with the induced metric $d_p(y,z) = |y-z|_p$. It is metrizable by the metric $d_p(y,z) := |y-z|_p^p$ when $0 < p < 1$. As the two-sided F-module $A_\infty(f)$ has the completion relative to the metric $d_p(\cdot,\cdot)$. Since F is complete relative to its norm, then the completion $A_{\infty,p}(f)$ of $A_\infty(f)$ is over F.

Derivations of Operator Algebras on Hypercomplex Hilbert Spaces ... 67

From formula (2.13), it follows that $|\bar{z}|_p = |z|_p$ and either $|T(z)|_p \le 2|z|_p$ when $1 \le p$ or $|T(z)|_p^p \le 2d_p(z,0)$ when $0 < p < 1$ for each $z \in A_{\infty,p}(f)$. Thus the trace map T is continuous from $A_{\infty,p}(f)$ into F.

Consider u_1 and u_2 in $A_{n+1}(f_{(n+1)})$. Certainly elements x_1, x_2, y_1 and y_2 in $A_n(f_{(n)})$ exist such that $u_j = x_j + y_j i_{2^n}$ for $j=1$ and $j=2$. From formulas (2.5), (2.13), and (2.15), it follows that

$$|u_1 u_2|_p^p \le |x_1 x_2 - f_{n+1}\bar{y}_2 y_1|_p^p + |y_2 x_1 + y_1 \bar{x}_2|_p^p$$
$$\le |x_1 x_2|_p^p + |f_{n+1}|^p |\bar{y}_2 y_1|_p^p + |y_2 x_1|_p^p + |y_1 \bar{x}_2|_p^p.$$

Let J_n denote the least positive constant so that $|x_1 x_2|_p^p \le J_n |x_1|_p^p |x_2|_p^p$ for each x_1 and x_2 in $A_n(f_{(n)})$. We have that

$$|u_j|_p^p = |x_j|_p^p + |y_j|_p^p, \text{ and}$$
$$|u_1|_p^p |u_2|_p^p = |x_1|_p^p |x_2|_p^p + |x_1|_p^p |y_2|_p^p + |x_2|_p^p |y_1|_p^p + |y_1|_p^p |y_2|_p^p.$$

If $|f_{n+1}| \le 1$, then $J_{n+1} \le J_n$. If $|f_{n+1}| > 1$, then $J_{n+1} \le |f_{n+1}|^p J_n$. From the condition $0 < \prod_{n,|f_n|\ge 1} |f_n| < \infty$ it follows that $\sup_n J_n =: J < \infty$. On the other hand, $|1z|_p = |1|_p |z|_p$ for each $z \in A_\infty(f)$, consequently, $1 \le J$. Therefore, by induction, we infer that

$$|u_1 u_2|_p \le J^{1/p} |u_1|_p |u_2|_p,$$

for each u_1 and u_2 in $A_\infty(f)$ and hence also in $A_{\infty,p}(f)$ since $A_\infty(f)$ is dense in its l_p completion $A_{\infty,p}(f)$. Thus the multiplication $A_{\infty,p}(f)^2 \ni (u_1, u_2) \mapsto u_1 u_2 \in A_{\infty,p}(f)$ is jointly continuous in u_1 and u_2. For $0 < p < \infty$ redefining the mapping

$$\|z\|_p = J^{1/p} |z|_p, \tag{2.16}$$

one gets that

$$\|u_1 u_2\|_p \le \|u_1\|_p \|u_2\|_p, \tag{2.17}$$

for each u_1 and u_2 in $A_{\infty,p}(f)$.

2. In this case, a norm on F is non-archimedean, that is,

$$|a+b| \le \max(|a|, |b|), \tag{2.18}$$

for each a and b in F. Next let
$$|z|_\infty := \sup_j |z_j|, \qquad (2.19)$$
for each $z \in A_\infty(f)$.

We remind that $c_0(F)$ is an F-module of all converging to zero sequences in F supplied with the norm given by formula (2.19). This gives the norm on $A_\infty(f)$. The completion of $A_\infty(f)$ as the F-module relative to this norm we denote by $A_{\infty,c_o}(f)$. Therefore, $|z|_\infty = |\bar{z}|_\infty$ for each $z \in A_{\infty,c_o}(f)$. This norm satisfies the strong triangle inequality
$$|y + z|_\infty \leq \max(|y|_\infty, |z|_\infty). \qquad (2.20)$$

Take arbitrary u_1 and u_2 in $A_{n+1}(f_{(n+1)})$, then for their product $w = u_1 u_2$, we infer that
$$|w|_\infty \leq \max(|x_1 x_2 - f_{n+1}\bar{y}_2 y_1|_\infty, |y_2 x_1 + y_1 \bar{x}_2|_\infty)$$
$$\leq \max(|x_1 x_2|_\infty, |f_{n+1}||\bar{y}_2 y_1|_\infty, |y_2 x_1|_\infty, |y_1 \bar{x}_2|_\infty),$$
where x_1, x_2, y_1 and y_2 are in $A_n(f_{(n)})$, $u_j = x_j + y_j i_{2^n}$ for $j = 1$ and $j = 2$.

Put K_n to be the least positive constant so that
$$|x_1 x_2|_\infty \leq K_n |x_1|_\infty |x_2|_\infty,$$
for each x_1 and x_2 in $A_n(f_{(n)})$. We mention that
$$|u_j|_\infty = \max(|x_j|_\infty, |y_j|_\infty), \text{ and}$$
$$|u_1|_\infty |u_2|_\infty \leq K_n \max(|x_1|_\infty |x_2|_\infty, |x_1|_\infty |y_2|_\infty, |x_2|_\infty |y_1|_\infty, |f_{n+1}||y_1|_\infty |y_2|_\infty).$$
When $|f_{n+1}| \leq 1$, the inequality is valid $K_{n+1} \leq K_n$. In the case $|f_{n+1}| > 1$, we deduce that
$$K_{n+1} \leq |f_{n+1}| K_n.$$
Using the condition $0 < \prod_{n, |f_n| \geq 1} |f_n| < \infty$, we infer that
$$\sup_n K_n =: K < \infty.$$
But $1 \leq K$ since $|1z|_\infty = |1|_\infty |z|_\infty$ for each $z \in A_\infty(f)$. For each u_1 and u_2 in $A_{\infty,c_o}(f)$, we have that $\lim_n u_{j,n} = 0$ for $j = 1$ and $j = 2$, where
$$u_j = \sum_{n=0}^{\infty} u_{j,n} i_n,$$

with $u_{j,n} \in F$, for each j and n.

Utilizing mathematical induction, we get that

$$|u_1 u_2|_\infty \leq K |u_1|_\infty |u_2|_\infty,$$

for each u_1 and u_2 in $A_\infty(f)$, consequently, in $A_{\infty,c_o}(f)$ also since $A_\infty(f)$ is dense in its c_0 completion $A_{\infty,c_o}(f)$. Therefore the multiplication $A_{\infty,c_o}(f)^2 \ni (u_1, u_2) \mapsto u_1 u_2 \in A_{\infty,c_o}(f)$ is jointly continuous in u_1 and u_2. Putting

$$\|z\|_\infty = K|z|_\infty, \tag{2.21}$$

one arrives to the conclusion that

$$\|u_1 u_2\|_\infty \leq \|u_1\|_\infty \|u_2\|_\infty, \tag{2.22}$$

for each u_1 and u_2 in $A_{\infty,c_o}(f)$.

3. Consider the particular case, when F is a field supplied with the multiplicative norm and F is complete relative to it. Then $A_{\infty,p}(f)$ and $A_{\infty,c_o}(f)$ are Banach spaces over F in the corresponding cases either **1** with $1 \leq p < \infty$ or **2**. From formulas (2.17) and (2.22) it follows that they are Banach algebras.

Remark 2.2. From the proof of Theorems 2.1, it also follows that the algebra $A_{\infty,2}(f)$ over $\mathbf{R}$ with $f_n = 1$ for each n has a structure of a hypercomplex Hilbert space and a Hilbert algebra.

Theorem 2.2. Let $A_n(f_{(n)})$ and $A_m(g_{(m)})$ be two generalized Cayley-Dickson algebras over an associative commutative and unital ring F with $\mathrm{char}(F) \neq 2$, where $2 \leq n \leq \infty$ and $2 \leq m \leq \infty$. Let also $h : A_n(f_{(n)}) \to A_m(g_{(m)})$ be an F-linear homomorphism. Then either $h(A_n(f_{(n)}))$ is isomorphic with $A_n(f_{(n)})$ or $h(A_n(f_{(n)})) = \{0\}$.

Proof. A mapping h is F linear and $h(yz) = h(y)h(z)$ for each y and z in $A_n(f_{(n)})$. Therefore,

$$h(1z) = h(1)h(z) = h(z), \text{ and}$$
$$h(z1) = h(z)h(1) = h(z),$$

for each $z \in A_n(f_{(n)})$. This implies that either $h(1) = 1$ or $h(1) = 0$. If $h(1) = 0$, then $h(A_n(f_{(n)})) = \{0\}$. If $h(1) = 1$, then $h(t) = t$ for each $t \in F$. That is the restriction of h to F is the identity mapping $h|_F = id$.

Thus it remains to consider the case $h|_F = id$. Since $\mathrm{char}(F) \neq 2$ these algebras $A_n(f_{(n)})$ and $A_m(g_{(m)})$ have bases over F. Take a basis $\{i_0, i_1, i_2, \cdots\}$ in $A_n(f_{(n)})$, where $i_0 = 1$. Relations $i_k i_j = t_{k,j,s} i_s$ and $i_k i_j = v_{k,j} i_j i_k$ for each $0 \leq j$ and $0 \leq k$, where $t_{k,j,s}$ and $v_{k,j}$ belong to $F \setminus \{0\}$, s is uniquely defined by j and k, imply that

$$h(i_k)h(i_j) = t_{k,j,s} h(i_s), \text{ and}$$
$$h(i_k)h(i_j) = v_{k,j} h(i_j)h(i_k),$$

for each j and k.

On the other hand, $i_j^* = -i_j$ for each $j \geq 1$, hence $h(\bar{z}) = \overline{h(z)}$ for each $z \in A_n(f_{(n)})$. Therefore,

$$h(T(yz)) = T(h(y)h(z)) = T(yz).$$

Consequently,

$$h(N(z)) = N(h(z)) = N(z),$$

for each y and z in $A_n(f_{(n)})$. We have $i_j^2 \in F \setminus \{0\}$ for each $j \geq 0$, hence $(h(i_j))^2 = i_j^2$ for all $j \geq 0$. Mention that $T(yz) = 0$ for each $y \in A_n(f_{(n)})$ if and only if $z = 0$. Together with $h(z) = z_0 i_0 + z_1 h(i_1) + z_2 h(i_2) + \cdots$ for each $z \in A_n(f_{(n)})$, where $z = z_0 i_0 + z_1 i_1 + z_2 i_2 + \cdots$ with $z_j \in F$ for each j, this leads to the conclusion that $h(A_n(f_{(n)}))$ is isomorphic with $A_n(f_{(n)})$.

Remark 2.3. Let F be a topological associative commutative and unital ring with $\mathrm{char}(F) \neq 2$ and $A_0 = F$ and let τ be a topology on a generalized Cayley-Dickson algebra $A_\infty(f)$ over F so that its restriction $\tau|_F$ to F coincides with an initial topology on F. Suppose that $A_\infty(f)$ is a topological algebra relative to such topology τ. By addition $A_\infty(f)$ is a topological commutative group and hence τ induces a uniformity which we denote by $\mathcal{V} = \mathcal{V}(\tau)$ (see also [39, 40]). We denote by $A_{\infty,\tau}(f)$ the completion of $A_\infty(f)$ relative to the uniformity $\mathcal{V}$.

Certainly for each finite $n \in \mathbf{N}$ a uniformity on a ring F induces a uniformity on a generalized Cayley-Dickson algebra $A_n(f_{(n)})$ over F. If F is complete relative to its uniformity, then $A_n(f_{(n)})$ is complete when $n \in \mathbf{N}$. Theorems 2.1 above provide examples of infinite dimensional generalized Cayley-Dickson topological algebras complete relative to their uniformities. In the purely algebraic case of the discrete topology τ_d one has $A_{\infty,\tau_d}(f) = A_\infty(f)$.

Theorem 2.3. Let $A = A_{\infty,\tau}(f)$ be a generalized Cayley-Dickson algebra (see Remark 2.3), let also h be a nontrivial F-linear homomorphism of A into A. Then h is an isomorphism of A with $h(A)$.

Proof. By the conditions of this theorem
$$h(ty) = th(y),$$
$$h(y+z) = h(y) + h(z), \text{ and}$$
$$h(yz) = h(y)h(z),$$

for every y and z in A and $t \in F$. In view of Remark 2.3, $A_\infty(f)$ is dense in A. On the other hand, by Theorem 2.2, $h(A_\infty)$ is isomorphic with A_∞ since h is nontrivial. Then
$$h(\bar{z}) = \overline{h(z)}, \text{ and}$$
$$h(T(yz)) = T(yz) = T(h(y)h(z)),$$

for each y and z in A. Therefore, $h(z) = 0$ if and only if $z = 0$ since $z = 0$ if and only if $T(yz) = 0$ for each $y \in A$. Thus h preserves the form $T(yz)$ and h is bijective. Therefore, there exists the inverse mapping $h^{-1} : h(A) \to A$. Evidently, h^{-1} is also a homomorphism from $h(A)$ onto A since $h^{-1}(h(A)) = A$. That is $h : A \to h(A)$ is an isomorphism.

Theorem 2.4. Let $A = A_{\infty,\tau}(f)$ be a generalized Cayley-Dickson algebra (see Remark 2.5) and let it be metrizable by a metric ρ such that there exists a fixed $v > 0$ for which $|x+y|^v \leq |x|^v + |y|^v$ and $|xy| \leq |x||y|$ and $|tx| \leq |t||x|$ for every $t \in F$ and x and y in A, where $\rho(x, 0) = |x|$. Let also h be a nontrivial F-linear homomorphism from A into A. Then for each $t \in F$ with $|t| > 1$ there exist a resolvent $(tI - h)^{-1}$ from A into A.

Proof. By virtue of Theorem 2.3, actually, h is a homeomorphism of A onto $h(A)$. Take arbitrary $M \in A \setminus \{0\}$ with $T(M) = 0$, then $M^2 \in F$ and hence $(h(M))^2 \in F$ since $M\bar{M} = -M^2$ and $\overline{M\bar{M}} = M\bar{M}$. Therefore,
$$|M^{2k}| = |(M^2)^k| = |((h(M))^2)^k| = |(h(M))^{2k}| \leq |M^2|^k,$$

and analogously
$$|M^{2k}| = |((h^{-1}(M))^2)^k| = |(h^{-1}(M))^{2k}| \leq |M^2|^k.$$

This implies that
$$\lim_{k\to\infty} |(h(M))^k|^{1/k} \le \sqrt{|M^2|}, \text{ and}$$
$$\lim_{k\to\infty} |(h^{-1}(M))^k|^{1/k} \le \sqrt{|M^2|},$$
consequently,
$$\lim_{k\to\infty} |(h(M))^k|^{1/k} = \sqrt{|M^2|}, \text{ and} \qquad (2.23)$$
$$\lim_{k\to\infty} |(h^{-1}(M))^k|^{1/k} = \sqrt{|M^2|},$$
since $|M^{2k+1}| \le |M^{2k}||M|$ and $\lim_k |M|^{1/k} = 1$, when $|M| > 0$.

At the same time, $|h(t1)| = |t1| = |t|$ for each $t \in F$. The homomorphism h of A has an extension to a homomorphism of a generalized Cayley-Dickson algebra B which is obtained from A by the doubling procedure. The topology on A induces the corresponding topology on $A \oplus Al$, where l denotes the doubling generator. Put $|zl| = |z|$ and $h(zl) = h(z)l$ for each $z \in A$. Mention that $T(zl) = 0$ for each $z \in A$. Therefore, the consideration above is applicable to the algebra B as well. Thus
$$\lim_{k\to\infty} |(h(z))^k|^{1/k} = \sqrt{|z^2|} \le |z|, \text{ and}$$
$$\lim_{k\to\infty} |(h^{-1}(z))^k|^{1/k} = \sqrt{|z^2|} \le |z|,$$
for each $z \in A$. Hence spectral radii $r(h) := \lim_k |h^k|^{1/k}$ and $r(h^{-1})$ of h and h^{-1} respectively are $r(h) \le 1$ and $r(h^{-1}) \le 1$, where $|h| := \sup_{|z|>0} |h(z)|/|z|$.

We have that $tI - h$ and $(tI - h^{-1})$ are linear over F operators from A into A and from $h(A)$ into A respectively and there exist their resolvents $(tI - h)^{-1}$ and $(tI - h^{-1})^{-1}$ for each $t \in F$ such that $|t| > 1$. Using the formulas $|x+y|^v \le |x|^v + |y|^v$ and $|xy| \le |x||y|$ and $|tx| \le |t||x|$ for every $t \in F$ and x and y in A, we deduce that they are given by formal power convergent series
$$(tI - h)^{-1} = \sum_{n=0}^{\infty} t^{-n-1} h^n, \text{ and}$$
$$(tI - h^{-1})^{-1} = \sum_{n=0}^{\infty} t^{-n-1} h^{-n},$$

where $h^n(z) = h(h^{n-1}(z))$ for each natural number $n \geq 2$, $h^1(z) = h(z)$ since $|(tI - h)^{-1}|^v \leq \sum_{n=0}^{\infty} |t|^{-(n+1)v} |h|^{nv} < \infty$.

Corollary 2.1. *If suppositions of Theorem 2.4 are fulfilled, then spectral radii satisfy the inequalities $r(h) \leq 1$ and $r(h^{-1}) \leq 1$.*

Theorem 2.5. *Suppose that $F = \mathbf{R}$, $A_0 = F$, $A = A_{\infty,2}(f)$. If h is a homomorphism from A into A, then h is continuous.*

Proof. If h is trivial there is nothing to prove. Suppose that h is nontrivial. By virtue of Theorem 2.3, h is an isomorphism of A with $h(A)$. Up to an isomorphism of Banach algebras, we can consider that $f_n \in \{-1, 1\}$ for each $n \in \mathbf{N}$.

Let at first $f_n = 1$ for each $n \in \mathbf{N}$. In the considered case $|z| = \sqrt{z\bar{z}} = \|z\|_2$ for each $z \in A$. We have that the real field $\mathbf{R}$ is the center of the Cayley-Dickson algebra A. Moreover, $z\bar{z} = \bar{z}z \in \mathbf{R}$ and hence $z^2\bar{z}^2 = (z\bar{z})^2$, consequently, $|z^2| = |z|^2$ for each $z \in A$. According to the proof of Theorem 2.4, h is an isomorphism from A onto $h(A)$ and $|h| = 1$ since h preserves the quadratic from $T(yz)$ for each y and z in A, $|z|^2 = N(z)$. Thus h is continuous.

Suppose now that $f_n = -1$ for each $n \in \mathbf{N}$. Then, we obtain the equality $\|z\|_2^2 = -N(z)$ for each $z \in A$ since $z = z_0 i_0 + z_1 i_1 + z_2 i_2 + \cdots$ with $z_n \in \mathbf{R}$ for each $n = 0, 1, 2, \cdots$ and $\|z\|_2^2 = z_0^2 + z_1^2 + z_2^2 + \cdots$. Therefore, the proof in this case is analogous to that of the first variant.

Let now

$$\Lambda_1 = \{n \in \mathbf{N} : f_n = 1\} \neq \emptyset, \text{ and}$$
$$\Lambda_2 = \{n \in \mathbf{N} : f_n = -1\} \neq \emptyset.$$

We have that in A the subalgebra $A_\infty(f)$ is dense. On the other hand,

$$A_{p,2}(1, 1, \cdots) \cap A_{q,2}(-1, -1, \cdots) = \mathbf{R}i_0, \qquad (2.24)$$

where p and q are the cardinalities of Λ_1 and Λ_2 correspondingly, where $i_0 = 1$. From (2.24), it follows that there are isometric embeddings of Banach algebras

$$g_1 : A_{p,2}(1, 1, \cdots) \hookrightarrow A, \text{ and}$$
$$g_2 : A_{q,2}(-1, -1, \cdots) \hookrightarrow A. \qquad (2.25)$$

Therefore, each element z of $A_\infty(f)$ is a finite sum of finite ordered products z_k of elements $x_{k,1}, x_{k,2}, \cdots$ belonging to $g_1(A_p(1, 1, \cdots))$ and of elements

$u_{k,1}, u_{k,2}, \cdots$ belonging to $g_2(A_{q,2}(-1,-1,\cdots))$, where $z_k \in \mathbf{R}i_k \setminus \{0\}$ and $x_{k,l} \in \mathbf{R}i_{m(k,l)} \setminus \{0\}$ and $u_{k,s} \in \mathbf{R}i_{m(k,s)} \setminus \{0\}$ for each k, l, s such that $T(z_k z_j) = 0$ for each $k \neq j$; $\{i_0, i_1, i_2, \cdots\}$ denotes an orthonormal basis of A such that $i_k i_s \in \mathbf{R}i_{n(k,s)}$ and $|i_k i_s| = |i_k||i_s|$ for each k, s, where $m(k,s)$ and $n(k,s)$ are in $\{0,1,2,\cdots\}$ for each k and s. Moreover, the restriction of h to $\mathbf{R}i_0$ is the identity mapping:

$$h|_{\mathbf{R}i_0} = id. \qquad (2.26)$$

According to the preceding two cases, we get that two restrictions $h|_{g_1(A_{p,2}(1,1,\cdots))}$ and $h|_{g_2(A_{q,2}(-1,-1,\cdots))}$ are continuous. Up to isomorphisms of Banach Cayley-Dickson algebras they are isometries,

$$|h|_{g_1(A_{p,2}(1,1,\cdots))}| = 1 \text{ and } |h|_{g_2(A_{q,2}(-1,-1,\cdots))}| = 1.$$

Therefore, from (2.24)-(2.26), it follows that $h: A_\infty(f) \to A$ is the continuous homomorphism since $|h(z_k)| \leq |z_k|$ and $T(z_k z_j) = 0$ for each $k \neq j$, consequently, $|h(z)| \leq |z|$ for each $z \in A_\infty(f)$. Thus, it has the continuous extension $v: A \to A$. Considering Cauchy (i.e., fundamental) sequences in A composed of elements belonging to $A_\infty(f)$ we deduce that $v = h$ on A.

Corollary 2.2. Suppose that $F = \mathbf{R}$, $A_0 = F$, $A = A_{\infty,2}(f)$. If h is a homomorphism from A onto A, then h is a continuous automorphism.

Proof. In view of Theorems 2.3 and 2.5, h is an isomorphism of A with $h(A)$. From the condition $h(A) = A$ of this Corollary it follows that h is an automorphism of A. Again, by Theorem 2.5, the automorphisms h and h^{-1} are continuous.

3. Derivations and Cohomologies of Nonassociative Hypercomplex Algebras and Hilbert Modules

Definition 3.1. Let G be a set with a single-valued binary operation (multiplication) $G^2 \ni (a,b) \mapsto ab \in G$ defined on G satisfying the conditions:

$$\text{for each } a \text{ and } b \text{ in } G \text{ there is a unique } x \in G \text{ with } ax = b, \qquad (3.1)$$

$$\text{and a unique } y \in G \text{ exists satisfying } ya = b, \qquad (3.2)$$

which are denoted by $x = a \backslash b = \mathrm{Div}_l(a,b)$ and $y = b/a = \mathrm{Div}_r(a,b)$ correspondingly, there exists a neutral (i.e., unit) element $e_G = e \in G$:

$$eg = ge = g \text{ for each } g \in G. \qquad (3.3)$$

The set of all elements $h \in G$ commuting and associating with G:

$$\mathrm{Com}(G) := \{a \in G : \forall b \in G, ab = ba\}, \qquad (3.4)$$
$$N_l(G) := \{a \in G : \forall b \in G, \forall c \in G, (ab)c = a(bc)\}, \qquad (3.5)$$
$$N_m(G) := \{a \in G : \forall b \in G, \forall c \in G, (ba)c = b(ac)\}, \qquad (3.6)$$
$$N_r(G) := \{a \in G : \forall b \in G, \forall c \in G, (bc)a = b(ca)\}, \qquad (3.7)$$
$$N(G) := N_l(G) \cap N_m(G) \cap N_r(G); \qquad (3.8)$$

$\mathcal{C}(G) := \mathrm{Com}(G) \cap N(G)$ is called the center $\mathcal{C}(G)$ of G.

We call G a metagroup if a set G possesses a single-valued binary operation and satisfies conditions (3.1)-(3.3) and

$$(ab)c = \mathsf{t}_3(a,b,c)a(bc), \qquad (3.9)$$

for each a, b and c in G, where $\mathsf{t}_3(a,b,c) \in F_\circ$, $F_\circ \subset \mathcal{C}(G)$; $F_\circ$ denotes a (proper or improper) subgroup of $\mathcal{C}(G)$, where t_3 shortens a notation $\mathsf{t}_{3,G}$.

Then G will be called a central metagroup if in addition to (3.9) it satisfies the condition:

$$ab = \mathsf{t}_2(a,b)ba, \qquad (3.10)$$

for each a and b in G, where $\mathsf{t}_2(a,b) \in F_\circ$. Particularly, $\mathrm{Inv}_l(a) = \mathrm{Div}_l(a,e)$ is a left inversion, $\mathrm{Inv}_r(a) = \mathrm{Div}_r(a,e)$ is a right inversion.

In view of the nonassociativity of G in general a product of several elements of G is specified as usually by opening "(" and closing ")" parentheses. For elements $a_1, \cdots, a_n$ in G we shall denote shortly by $\{a_1, \cdots, a_n\}_{q(n)}$ the product, where a vector $q(n)$ indicates an order of pairwise multiplications of elements in the row $a_1, \cdots, a_n$ in braces in the following manner. Enumerate positions: before a_1 by 1, between a_1 and a_2 by 2,$\cdots$, by n between a_{n-1} and a_n, by $n+1$ after a_n. Then put $q_j(n) = (k,m)$ if there are k opening "(" and m closing ")" parentheses in the ordered product at the j-th position of the type $)\cdots)(\cdots($, where k and m are nonnegative integers, $q(n) = (q_1(n), \cdots, q_{n+1}(n))$ with $q_1(n) = (k,0)$ and $q_{n+1}(n) = (0,m)$.

As traditionally S_n denotes the symmetric group of the set $\{1, 2, \cdots, n\}$. Henceforth, maps and functions on metagroups are supposed to be single-valued if something others will not be specified.

Let $\psi : G \to G$ be a bijective surjective map satisfying the following condition: $\psi(ab) = \psi(a)\psi(b)$ for each a and b in G. Then ψ is called an automorphism of the metagroup G.

Lemma 3.1.

1. Let G be a central metagroup. Then for every $a_1, \cdots, a_n$ in G, $v \in S_n$ and vectors $q(n)$ and $u(n)$ indicating an order of pairwise multiplications and $n \in \mathbf{N}$ there exists an element $t_n = t_n(a_1, \cdots, a_n; q(n), u(n)|v) \in F_\circ$ such that

$$\{a_1, \cdots, a_n\}_{q(n)} = t_n\{a_{v(1)}, \cdots, a_{v(n)}\}_{u(n)}. \qquad (3.11)$$

2. If G is a metagroup and if v is the neutral element $v = id$ in S_n, then property (3.11) is satisfied.

Proof. From conditions (3.1)-(3.8), it follows that $\mathcal{C}(G)$ itself is a commutative group.

1. For $n = 1$, evidently, $t_1 = 1$ since $a = 1a$ for each $a \in G$. For $n = 2$, formula (3.11) is a direct consequence of condition (3.10). Consider $n = 3$. When u is the identity element of S_3 the statement follows from condition (3.9). For any transposition u of two elements of the set $\{1, 2, 3\}$ the statement follows from (3.9) and (3.10). Elements of S_3 can be obtained by multiplication of pairwise transpositions. Therefore from the condition $F_\circ \subset \mathcal{C}(G)$ it follows that formula (3.11) is valid.

Let now $n \geq 4$ and suppose that this Lemma is proved for any products consisting of less than n elements. In view of properties (3.1) and (3.2) it is sufficient to verify formula (3.11) for

$$\{a_1, \cdots, a_n\}_{q(n)} = (\cdots((a_1 a_2)a_3)\cdots)a_n =: \{a_1, \cdots, a_n\}_{l(n)},$$

since $F_\circ \subset \mathcal{C}(G)$. In the particular case:

$$\{a_{v(1)}, \cdots, a_{v(n)}\}_{u(n)} = \{a_{v(1)}, \cdots, a_{v(n-1)}\}_{u(n-1)} a_n,$$

Derivations of Operator Algebras on Hypercomplex Hilbert Spaces ...

formula (3.11) follows from the induction hypothesis since

$$(\cdots((a_1a_2)a_3)\cdots)a_{n-1} = t_{n-1}\{a_{v(1)},\cdots,a_{v(n-1)}\}_{u(n-1)}$$

and hence

$$((\cdots((a_1a_2)a_3)\cdots)a_{n-1})a_n = t_{n-1}(\{a_{v(1)},\cdots,a_{v(n-1)}\}_{u(n-1)}a_n)$$

and putting $t_n = t_{n-1}$, where $t_{n-1} = t_{n-1}(a_1,\cdots,a_{n-1};q(n-1),u(n-1)|w)$ with $w = v|_{\{1,\cdots,n-1\}}$, $v(n) = n$.

In the general case,

$$\{a_{v(1)},\cdots,a_{v(n)}\}_{u(n)} = \{b_1,\cdots,b_j,\cdots,b_k\}_{p(k)},$$

where j is such that either $b_j = c_j a_n$ with $c_j = \{a_{v(j)},\cdots,a_{v(j+m-1)}\}_{r(m)}$ and with $v(j+m) = n$ or $b_j = a_n c_j$ with $c_j = \{a_{v(j+1)},\cdots,a_{v(j+m)}\}_{r(m)}$ and with $v(j) = n$, also $b_1 = a_{v(1)},\cdots,b_{j-1} = a_{v(j-1)}$, $b_{j+1} = a_{v(j+1)},\cdots$, $b_k = a_{v(n)}$ with suitable vectors $p(k)$ and $r(m)$. If $m > 1$, then $k < n$ and using the induction hypothesis for $\{b_1,\cdots,b_j,\cdots,b_k\}_{p(k)}$ and b_j, we get that elements s and t in $F_\circ$ exist so that

$$\{b_1,\cdots,b_j,\cdots,b_k\}_{p(k)} = s\{b_1,\cdots,b_{j-1},b_{j+1},,\cdots,b_k\}_{p(k-1)}b_j$$
$$= st(\{b_1,\cdots,b_{j-1},b_{j+1},,\cdots,b_k\}_{p(k-1)}c_j)a_n,$$

where $p(k-1)$ is a corresponding vector prescribing an order of multiplications. Again, applying the induction hypothesis to the product of $n-1$ elements $\{b_1,\cdots,b_{j-1},b_{j+1},,\cdots,b_k\}_{p(k-1)}c_j$ we deduce that there exists $w \in F_\circ$ such that

$$\{a_{v(1)},\cdots,a_{v(n)}\}_{u(n)} = stw((\cdots((a_1a_2)a_3)\cdots)a_{n-1})a_n.$$

Therefore, a case remains when $m = 1$. Let the first multiplication in $\{a_{v(1)},\cdots,a_{v(n)}\}_{u(n)}$ containing a_n be $(a_{v(k)}a_{v(k+1)}) =: b_k$, consequently,

$$\{a_{v(1)},\cdots,a_{v(n)}\}_{u(n)} = \{b_{y(1)},\cdots,b_{y(n-1)}\}_{w(n-1)},$$

for some $y \in S_{n-1}$ and a vector $w(n-1)$ indicating an order of pairwise products, where $b_j = a_{v(j)}$ for each $1 \le j \le k-1$, also $b_{j-1} = a_{v(j)}$ for each $k+1 < j \le n$, where either $a_n = a_{v(k)}$ or $a_n = a_{v(k+1)}$. From the induction hypothesis we deduce that there exists $t_{n-1} \in F_\circ$ so that

$$t_{n-1}\{b_{y(1)},\cdots,b_{y(n-1)}\}_{w(n-1)} = pb_k$$

with $p = \{b_1, \cdots, b_{k-1}, b_{k+1}, \cdots, b_{n-1}\}_{w(n-1)}$. Applying the induction hypothesis for $n = 3$ we infer that there exists $t_3 \in F_o$ such that

$$t_{n-1}t_3\{b_{y(1)}, \cdots, b_{y(n-1)}\}_{w(n-1)} = (pa)a_n,$$

where either $a = a_{v(k+1)}$ or $a = a_{v(k)}$ correspondingly. From the induction hypothesis for $n - 1$ it follows that there exists $\tilde{t}_{n-1} \in F_o$ so that

$$\tilde{t}_{n-1}pa = (\cdots ((a_1 a_2)a_3) \cdots)a_{n-1}$$

and hence

$$\{a_1, \cdots, a_n\}_{l(n)} = t_n\{a_{v(1)}, \cdots, a_{v(n)}\}_{u(n)},$$

where $t_n = \tilde{t}_{n-1}t_{n-1}t_3$.

2. Let now G be a metagroup and $v = id$ be the neutral element of the symmetric group S_n, where $id(k) = k$ for each $k \in \mathbf{N}$. Then, condition (3.10) is unnecessary because transpositions are already not used. For $n = 1$ and $n = 2$ we get $t_1 = 1$ and $t_2 = 1$ since $a = 1a$ and $ab = 1ab$ for each a and b in G. For $n = 3$, formula (3.11) follows from condition (3.9). Then, the proof in the case **2** by induction is a simplification of that of the case **1**.

Lemma 3.2. If G is a metagroup, then for each a and $b \in G$ the following identities are fulfilled:

$$b \backslash e = (e/b) t_3(e/b, b, b \backslash e); \qquad (3.12)$$
$$(a \backslash e)b = (a \backslash b) t_3(e/a, a, a \backslash e)/t_3(e/a, a, a \backslash b); \qquad (3.13)$$
$$b(e/a) = (b/a) t_3(b/a, a, a \backslash e)/t_3(e/a, a, a \backslash e). \qquad (3.14)$$

Proof. Conditions (3.1)-(3.3) imply that

$$b(b \backslash a) = a, \; b \backslash (ba) = a; \qquad (3.15)$$
$$(a/b)b = a, \; (ab)/b = a, \qquad (3.16)$$

for each a and b in G. Using condition (3.9) and identities (3.15) and (3.16), we deduce that

$$e/b = (e/b)(b(b \backslash e)) = (b \backslash e)/t_3(e/b, b, b \backslash e),$$

which leads to (3.12).

Let $c = a \setminus b$, then from identities (3.12) and (3.15) it follows that
$$(a\setminus e)b = (e/a)\mathsf{t}_3(e/a, a, a\setminus e)(ac) = ((e/a)a)(a\setminus b)\mathsf{t}_3(e/a, a, a\setminus e)/\mathsf{t}_3(e/a, a, a\setminus b),$$
which provides (3.13).

Let now $d = b/a$, then identities (3.12) and (3.16) imply that
$$b(e/a) = (da)(a\setminus e)/\mathsf{t}_3(e/a, a, a\setminus e) = (b/a)\mathsf{t}_3(b/a, a, a\setminus e)/\mathsf{t}_3(e/a, a, a\setminus e),$$
which demonstrates (3.14).

Definition 3.2. Let A be an algebra over an associative unital ring $\mathcal{T}$ such that A has a natural structure of a $(\mathcal{T}, \mathcal{T})$-bimodule with a multiplication map $A \times A \to A$, which is right and left distributive $a(b + c) = ab + ac$, $(b + c)a = ba + ca$, also satisfying the following identities $r(ab) = (ra)b$, $(ar)b = a(rb)$, $(ab)r = a(br)$, $s(ra) = (sr)a$ and $(ar)s = a(rs)$ for any a, b and c in A, r and s in $\mathcal{T}$. Let G be a metagroup and $\mathcal{T}$ be an associative unital ring satisfying conditions (3.17)-(3.19):

$$sa = as \text{ for each } a \in G, \ s \in \mathcal{T}, \tag{3.17}$$

$$s(ra) = (sr)a \text{ for each } s \text{ and } r \text{ in } \mathcal{T}, \text{ and } a \in G, \tag{3.18}$$

$$r(ab) = (ra)b, \ (ar)b = a(rb), \ (ab)r = a(br), \tag{3.19}$$

for each a and b in G, $r \in \mathcal{T}$.

Then by $\mathcal{T}[G]$ is denoted a metagroup algebra over $\mathcal{T}$ of all formal sums $s_1 a_1 + \cdots + s_n a_n$, where $s_1, \cdots, s_n$ are in $\mathcal{T}$ and $a_1, \cdots, a_n$ belong to G. Henceforth the ring $\mathcal{T}$ will be supposed commutative and associative, if something other will not be specified.

Note. Let M be an additive commutative group such that M is a two-sided G-module, where G is a metagroup. Remind that this means that to each $g \in G$ there correspond automorphisms $p(g)$ and $s(g)$ of M. We put for short $gx = p(g)x$ and $xg = xs(g)$ for each $g \in G$.

Evidently, M is a two-sided G-module if and only if it is a two-sided $\mathbf{Z}[G]$-module according to the formulas

$$\left(\sum_{g \in G} n(g)g\right)x = \sum_{g \in G} n(g)(gx), \text{ and}$$

$$x\left(\sum_{g \in G} n(g)g\right) = \sum_{g \in G} (xg)n(g),$$

where $n(g) \in \mathbf{Z}$ for each $g \in G$, $\mathbf{Z}$ denotes the ring of all integers.

One can consider the additive group of integers $\mathbf{Z}$ as the trivial two-sided G-module putting $gn = ng = n$ for each $g \in G$ and $n \in \mathbf{Z}$, where G is a metagroup.

Naturally, there may be a case when a nonassociative ring Y exists such that T and G are contained in Y. It may happen that G and T have common elements. This may induce the corresponding relations in $A = T[G]$. Another particular case is when T and G have no any common elements.

Example 3.1. Recall the following. Let A be a unital algebra over a commutative associative unital ring F supplied with a scalar involution $a \mapsto \bar{a}$ so that its norm N and trace T maps have values in F and fulfil conditions:

$$a\bar{a} = N(a)1 \text{ with } N(a) \in F, \tag{3.20}$$
$$a + \bar{a} = T(a)1 \text{ with } T(a) \in F, \tag{3.21}$$
$$T(ab) = T(ba), \tag{3.22}$$

for each a and b in A.

If a scalar $f \in F$ satisfies the condition: $\forall a \in A$, $fa = 0 \Rightarrow a = 0$, then such element f is called cancelable. For a cancelable scalar f the Cayley-Dickson doubling procedure provides new algebra $C(A, f)$ over F such that:

$$C(A, f) = A \oplus Al, \tag{3.23}$$
$$(a + bl)(c + dl) = (ac - f\bar{d}b) + (da + b\bar{c})l \tag{3.24}$$
$$\text{and } \overline{(a + bl)} = \bar{a} - bl, \tag{3.25}$$

for each a and b in A. Then l is called a doubling generator. From definitions of T and N it follows that $\forall a, b \in A$,

$$T(a) = T(a + bl), \text{ and}$$
$$N(a + bl) = N(a) + fN(b).$$

The algebra A is embedded into $C(A, f)$ as $A \ni a \mapsto (a, 0)$, where $(a, b) = a + bl$. Put by induction $A_n(f_{(n)}) = C(A_{n-1}, f_n)$, where $A_0 = A$, $f_1 = f$, $n = 1, 2, \cdots$, $f_{(n)} = (f_1, \cdots, f_n)$. Then $A_n(f_{(n)})$ are generalized Cayley-Dickson algebras when F is not a field, or Cayley-Dickson algebras when F is a field.

It is natural to put
$$A_\infty(f) := \bigcup_{n=1}^{\infty} A_n(f_{(n)}),$$
where $f = (f_n : n \in \mathbf{N})$. If $\mathrm{char}(F) \neq 2$, let $\mathrm{Im}(z) = z - T(z)/2$ be the imaginary part of a Cayley-Dickson number z and hence
$$N(a) := N_2(a, \bar{a})/2,$$
where $N_2(a, b) := T(a\bar{b})$.

If the doubling procedure starts from $A = F1 =: A_0$, then $A_1 = C(A, f_1)$ is a $*$-extension of F. If A_1 has a basis $\{1, u\}$ over F with the multiplication table $u^2 = u + w$, where $w \in F$ and $4w + 1 \neq 0$, with the involution $\bar{1} = 1$, $\bar{u} = 1 - u$, then A_2 is the generalized quaternion algebra, A_3 is the generalized octonion (Cayley-Dickson) algebra.

When $F = \mathbf{R}$ and $f_n = 1$ for each n by $\mathcal{A}_r$ will be denoted the real Cayley-Dickson algebra with generators $i_0, \cdots, i_{2^r-1}$ such that $i_0 = 1$, $i_j^2 = -1$ for each $j \geq 1$, $i_j i_k = -i_k i_j$ for each $j \neq k \geq 1$. Frequently $\bar{a}$ is also denoted by a^* or $\tilde{a}$.

Let A_n be a Cayley-Dickson algebra over a commutative associative unital ring $\mathcal{R}$ of characteristic different from two such that $A_0 = \mathcal{R}$, $n \geq 2$. Take its basic generators $i_0, i_1, \cdots, i_{2^n-1}$, where $i_0 = 1$. Choose $F_\circ$ as a multiplicative subgroup contained in the ring $\mathcal{R}$ such that $f_j \in F_\circ$ for each $j = 0, \cdots, n$. Put $G_n = \{i_0, i_1, \cdots, i_{2^n-1}\} \times F_\circ$. Then G_n is a central metagroup.

More generally, let H be a group such that $F_\circ \subset H$ with relations $h i_k = i_k h$ and $(hg)i_k = h(g i_k)$ for each $k = 0, 1, \cdots, 2^n - 1$ and each h and g in H. Then $G_n = \{i_0, i_1, \cdots, i_{2^n-1}\} \times H$ is also a metagroup. The latter metagroup is noncentral when H is noncommutative. Analogously to the Cayley-Dickson algebra $A_\infty(f)$, a metagroup G_∞ corresponds to $n = \infty$.

Generally, metagroups need not be central. From given metagroups new metagroups can be constructed using their direct or semidirect products. Certainly each group is a metagroup also. Therefore, there are abundant families of noncentral metagroups and also of central metagroups different from groups.

In another way smashed products of groups and of metagroups can be considered providing another examples of metagroups (see in more details in Appendix.)

Definition 3.3. Let $\mathcal{R}$ be a ring, which may be nonassociative relative to the multiplication. If there exists a mapping $\mathcal{R} \times M \to M$, $\mathcal{R} \times M \ni (a, m) \mapsto am \in M$ such that

$$a(m + k) = am + ak, \text{ and}$$
$$(a + b)m = am + bm, \quad (3.26)$$

for each a and b in $\mathcal{R}$, m and k in M, then M will be called a generalized left $\mathcal{R}$-module or shortly: left $\mathcal{R}$-module or left module over $\mathcal{R}$.

If $\mathcal{R}$ is a unital ring and $1m = m$ for each $m \in M$, then M is called a left unital module over $\mathcal{R}$, where 1 denotes the unit element in the ring $\mathcal{R}$. Symmetrically is defined a right $\mathcal{R}$-module.

If M is a left and right $\mathcal{R}$-module, then it is called a two-sided $\mathcal{R}$-module or a $(\mathcal{R}, \mathcal{R})$-bimodule. If M is a left $\mathcal{R}$-module and a right $\mathcal{S}$-module, then it is called a $(\mathcal{R}, \mathcal{S})$-bimodule.

A two-sided module M over $\mathcal{R}$ is called cyclic if an element $y \in M$ exists such that

$$M = \mathcal{R}(y\mathcal{R}) = \{s(yp) : s, p \in \mathcal{R}\}, \text{ and}$$
$$M = (\mathcal{R}y)\mathcal{R} = \{(sy)p : s, p \in \mathcal{R}\}.$$

Taking a metagroup algebra $A = \mathcal{T}[G]$ and a two-sided A-module M, where $\mathcal{T}$ is an associative unital ring satisfying conditions (3.17)-(3.19). Let M has the decomposition $M = \sum_{g \in G} M_g$ as a two-sided $\mathcal{T}$-module, where M_g is a two-sided $\mathcal{T}$-module for each $g \in G$, G is a metagroup, and let M satisfies the following conditions:

$$hM_g = M_{hg} \text{ and } M_g h = M_{gh}, \quad (3.27)$$
$$(bh)x_g = b(hx_g) \text{ and } x_g(bh) = (x_g h)b \text{ and } bx_g = x_g b, \quad (3.28)$$
$$(hs)x_g = \mathsf{t}_3(h, s, g)h(sx_g) \text{ and } (hx_g)s = \mathsf{t}_3(h, g, s)h(x_g s)$$
$$\text{and } (x_g h)s = \mathsf{t}_3(g, h, s)x_g(hs), \quad (3.29)$$
$$(bc)x = b(cx), \ (bx)c = b(xc), \ (xb)c = x(bc), \quad (3.30)$$

for every h, g, s in G, b and c in $\mathcal{T}$, $x \in M$ and $x_g \in M_g$. Then a two-sided A-module M satisfying (3.27)-(3.30) will be called smashly G-graded. Shortly, we shall say that M is G-graded. In particular, if a sum above is direct $M = \bigoplus_{g \in G} M_g$, then it will be said that M is directly G-graded.

Similarly are defined G-graded left and right A-modules. It will be said shortly an A-module instead of a G-graded A-module, when the metagroup G is given.

If P and N are left A-modules and a homomorphism $\gamma: P \to N$ is such that $\gamma(ax) = a\gamma(x)$ for each $a \in A$ and $x \in P$, then γ is called a left A-homomorphism. Analogously are defined right A-homomorphisms for right A-modules. For two-sided A modules a left and right A-homomorphism is called an A-homomorphism.

For left $\mathcal{T}$-modules M and N by $\mathrm{Hom}_{\mathcal{T}}(M, N)$ is denoted a family of all left $\mathcal{T}$-homomorphisms from M into N. A similar notation is used for a family of all $\mathcal{T}$-homomorphisms (or right $\mathcal{T}$-homomorphisms) of two-sided $\mathcal{T}$-modules (or right $\mathcal{T}$-modules correspondingly). If a ring $\mathcal{R}$ is specified it may be written shortly a homomorphism instead of a $\mathcal{R}$-homomorphism.

Example 3.2. Let $\mathcal{T}$ be a commutative associative unital ring, let also G be a metagroup and $A = \mathcal{T}[G]$ be a metagroup algebra (see Definition 3.2), where A is considered as a $\mathcal{T}$-algebra. Put $K_{-1} = A$, $K_0 = A \otimes_{\mathcal{T}} A$ and by induction $K_{n+1} = K_n \otimes_{\mathcal{T}} A$ for each natural number n. Each K_n is supplied with a two-sided A-module structure, $\forall p \in \mathcal{T}[\mathsf{C}(G)]$,

$$p \cdot (x_0, \cdots, x_{n+1}) = ((px_0), \cdots, x_{n+1}), \text{ and}$$
$$(x_0, \cdots, (x_{n+1}p)) = (x_0, \cdots, x_{n+1}) \cdot p, \qquad (3.31)$$

and $\forall j \in \{1, \cdots, n\}$,

$$p \cdot (x_0, \cdots, x_{n+1}) = (x_0, \cdots, (px_j), \cdots, x_{n+1}), \text{ and}$$
$$(x_0, \cdots, (x_j p), \cdots, x_{n+1}) = (x_0, \cdots, x_{n+1}) \cdot p, \qquad (3.32)$$

where $0 \cdot (x_1, \cdots, x_n) = 0$;

$$(xy) \cdot (x_0, \cdots, x_{n+1}) = \mathsf{t}_3 \cdot (x \cdot (y \cdot (x_0, \cdots, x_{n+1}))) \qquad (3.33)$$

with $\mathsf{t}_3 = \mathsf{t}_3(x, y, b)$, (see also formula (3.9) above);

$$\mathsf{t}_3 \cdot ((x_0, \cdots, x_{n+1}) \cdot (xy)) = ((x_0, \cdots, x_{n+1}) \cdot x) \cdot y \text{ with } \mathsf{t}_3 = \mathsf{t}_3(b, x, y); \qquad (3.34)$$
$$(x \cdot (x_0, \cdots, x_{n+1})) \cdot y = \mathsf{t}_3 \cdot (x \cdot ((x_0, \cdots, x_{n+1}) \cdot y)) \text{ with } \mathsf{t}_3 = \mathsf{t}_3(x, b, y) \qquad (3.35)$$

$$x \cdot (x_0, \cdots, x_{n+1}) = t_{n+3}(x, x_0, \cdots, x_{n+1}; v_0(n+3); l(n+3)) \cdot ((xx_0), x_1, \cdots, x_{n+1}), \qquad (3.36)$$

where
$$\{x, x_0, \cdots, x_{n+1}\}_{v_0(n+3)} = x\{x_0, \cdots, x_{n+1}\}_{l(n+2)},$$
$$\{x_0, \cdots, x_{n+1}\}_{l(n+2)} = \{x_0, \cdots, x_n\}_{l(n+1)} x_{n+1},$$
$$\{x_0\}_{l(1)} = x_0, \{x_0 x_1\}_{l(2)} = x_0 x_1;$$

where
$$b = \{x_0, \cdots, x_{n+1}\}_{l(n+2)},$$
$$t_n(x_1, \cdots, x_n; u(n), w(n)) := t_n(x_1, \cdots, x_n; u(n), w(n)|id).$$

Using shortened notation;

$$(x_0, \cdots, x_{n+1}) \cdot x =$$
$$t_{n+3}(x_0, \cdots, x_{n+1}, x; l(n+3), v_{n+2}(n+3)) \cdot (x_0, \cdots, x_n, (x_{n+1}x)) \tag{3.37}$$

for every $x, y, x_0, \cdots, x_{n+1}$ in G, where $(x_0, \cdots, x_{n+1})$ denotes an element of K_n over $\mathcal{T}$ corresponding to the left ordered tensor product $(\cdots((x_0 \otimes x_1) \otimes x_2) \cdots \otimes x_n) \otimes x_{n+1}$, $\{x_0, \cdots, x_{n+1}, x\}_{v_{n+2}(n+3)} = \{x_0, \cdots, x_n, x_{n+1}x\}_{l(n+2)}$.

Elements of the form $\{x_1, \cdots, x_n\}_{q(n)}$ will be called generating, where $x_1, \cdots, x_n$ belong to G (see also Definition 3.1).

Proposition 3.1. For each metagroup algebra $A = \mathcal{T}[G]$ (see Definition 3.2) an acyclic left A-complex $\mathcal{K}$ exists.

Proof. Take two-sided A-modules K_n as in the Example 3.2. There exists a boundary $\mathcal{T}$-linear operator $\partial_n : K_n \to K_{n-1}$ on K_n. On generating elements it will be given by the formulas:

$$\partial_n((x \cdot (x_0, x_1, \cdots, x_n, x_{n+1})) \cdot y) =$$
$$\sum_{j=0}^{n} (-1)^j t_{n+4}(x, x_0, \cdots, x_{n+1}, y; l(n+4), u_{j+1}(n+4)) \cdot$$
$$((x \cdot (\langle x_0, x_1, \cdots, x_{n+1}\rangle_{j+1, n+2})) \cdot y), \tag{3.38}$$

where

$$\langle x_0, \cdots, x_{n+1}\rangle_{1,n+2} := ((x_0 x_1), x_2, \cdots, x_{n+1}), \tag{3.39}$$

$$\langle x_0, \cdots, x_{n+1}\rangle_{2,n+2} := (x_0, (x_1 x_2), x_3, \cdots, x_{n+1}), \cdots, \tag{3.40}$$

$$\langle x_0, \cdots, x_{n+1}\rangle_{n+1,n+2} := (x_0, \cdots, x_{n-1}, (x_n x_{n+1})), \tag{3.41}$$

$$\partial_0(x \cdot (x_0, x_1)) \cdot y = (x \cdot (x_0 x_1)) \cdot y, \tag{3.42}$$

$$\{x_0, x_1, \cdots, x_{n+1}\}_{l(n+2)} := (\cdots((x_0 x_1) x_2) \cdots) x_{n+1}; \tag{3.43}$$

$$\{x, x_0, \cdots, x_{n+1}, y\}_{u_1(n+4)} := (x\{(x_0 x_1), x_2, \cdots, x_{n+1}\}_{l(n+1)}) y, \cdots, \tag{3.44}$$

$$\{x, x_0, \cdots, x_{n+1}, y\}_{u_{n+1}(n+4)} := (x\{x_0, x_1, \cdots, (x_n x_{n+1})\}_{l(n+1)}) y, \tag{3.45}$$

for each $x, x_0, \cdots, x_{n+1}, y$ in G. On the other hand, from formulas (3.1) and (3.2) it follows that

$$t_{n+4}(x, x_0, \cdots, x_{n+1}, y; l(n+4), u_{j+1}(n+4)) =$$

$$t_{n+2}(x_0, \cdots, x_{n+1}; l(n+2), v_{j+1}(n+2))$$

for each $j = 0, \cdots, n$, where

$$\{x_0, \cdots, x_{n+1}\}_{v_1(n+2)} := \{(x_0 x_1), x_2, \cdots, x_{n+1}\}_{l(n+1)}, \cdots, \tag{3.46}$$

$$\{x_0, \cdots, x_{n+1}\}_{v_{n+1}(n+2)} := \{x_0, x_1, \cdots, (x_n x_{n+1})\}_{l(n+1)}, \tag{3.47}$$

for every $x_0, \cdots, x_{n+1}$ in G. Therefore, ∂_n is a left and right A-homomorphism of (A, A)-modules. In particular,

$$\partial_1((x \cdot (x_0, x_1, x_2)) \cdot y = (x \cdot ((x_0 x_1), x_2)) \cdot y - t_3(x_0, x_1, x_2) \cdot (x \cdot (x_0, (x_1 x_2))) \cdot y,$$
$$\partial_2((x \cdot (x_0, x_1, x_2, x_3)) \cdot y = (x \cdot ((x_0 x_1), x_2, x_3)) \cdot y - t_4(x_0, \cdots, x_3; l(4), v_2(4)) \cdot$$
$$((x \cdot (x_0, (x_1 x_2), x_3)) \cdot y) + t_4(x_0, \cdots, x_3; l(4), v_3(4); id)((x \cdot (x_0, x_1, (x_2 x_3))) \cdot y).$$

Define a $\mathcal{T}$-linear homomorphism $s_n : K_n \to K_{n+1}$, which on generating elements has the form:

$$s_n(x_0, \cdots, x_{n+1}) = (1, x_0, \cdots, x_{n+1}), \tag{3.48}$$

for every $x_0, \cdots, x_{n+1}$ in G. From formulas (3.8), (3.9), and (3.11) the identities

$$t_n(x_1, \cdots, x_n; q(n), u(n)|v)t_n(x_1, \cdots, x_n; u(n), q(n)|v^{-1}) = 1, \quad (3.49)$$

$$t_n(x_1, \cdots, x_n; q(n), u(n))t_n(x_1, \cdots, x_n; u(n), w(n)) = t_n(x_1, \cdots, x_n; q(n), w(n)), \quad (3.50)$$

follow for every elements $x_1, \cdots, x_n$ in the metagroup G, vectors $q(n)$, $u(n)$, and $w(n)$ indicating orders of their multiplications, $v \in S_n$ and $n \in \mathbf{N}$. The following identity is evident:

$$t_{n+1}(1, x_1, \cdots, x_n; q(n+1), u(n+1)|v(n+1)) = t_n(x_1, \cdots, x_n; q(n), u(n)|v(n)) \quad (3.51)$$

for data $q(n)$, $u(n)$, and $v(n)$ obtained from $q(n+1)$, $u(n+1)$, and $v(n+1)$ correspondingly by taking into account the identity $1b = b1 = b$ for each $b \in G$. Hence

$$s_n((x_0, \cdots, x_{n+1}) \cdot y) = (s_n(x_0, \cdots, x_{n+1})) \cdot y,$$

for every $x_0, \cdots, x_{n+1}, y$ in G.

Let $p_n : K_{n+1} \to K_n$ be a $\mathcal{T}$-linear mapping such that

$$p_n(a \otimes b) = a \cdot b, \text{ and } p_n(b \otimes a) = b \cdot a, \quad (3.52)$$

for each $a \in K_n$ and $b \in A$. Therefore, from formulas (3.50) and (3.51) we deduce that $p_n s_n = id$ is the identity on K_n, consequently, s_n is a monomorphism.

Therefore, from formulas (3.17)-(3.19), (3.27)-(3.30), (3.38) and (3.46), (3.48), (3.50), (3.51) we infer that

$$(\partial_{n+1} s_n + s_{n-1} \partial_n)(x_0, \cdots, x_{n+1})$$
$$= \partial_{n+1}(1, x_0, \cdots, x_{n+1})$$
$$+ s_{n-1}\left(\sum_{j=0}^{n}(-1)^j t_{n+2}(x_0, \cdots, x_{n+1}; l(n+2), v_{j+1}(n+2)) \cdot \langle x_0, \cdots, x_{n+1}\rangle_{j+1,n+2}\right)$$
$$= \sum_{j=0}^{n+1}(-1)^j t_{n+3}(1, x_0, \cdots, x_{n+1}; l(n+3), v_{j+1}(n+3)) \cdot \langle 1, x_0, x_1, \cdots, x_{n+1}\rangle_{j+1,n+3}$$
$$+ \sum_{j=0}^{n}(-1)^j t_{n+2}(x_0, \cdots, x_{n+1}; l(n+2), v_{j+1}(n+2)) \cdot \langle 1, x_0, \cdots, x_{n+1}\rangle_{j+2,n+3}$$
$$= (x_0, \cdots, x_{n+1}).$$

for every $x_0, \cdots, x_{n+1}$ in G.

Thus the homotopy conditions

$$\partial_{n+1} s_n + s_{n-1}\partial_n = 1 \text{ for each } n \geq 0, \tag{3.53}$$

are fulfilled, where 1 denotes the identity operator on K_n. Therefore the recurrence relation

$$\partial_n \partial_{n+1} s_n = s_{n-2}\partial_{n-1}\partial_n, \tag{3.54}$$

is accomplished since

$$\partial_n \partial_{n+1} s_n = \partial_n(1 - s_{n-1}\partial_n) = \partial_n - (\partial_n s_{n-1})\partial_n = \partial_n - (1 - s_{n-2}\partial_{n-1})\partial_n.$$

On the other hand, from formula (3.48) it follows that K_{n+1} as the left A-module is generated by $s_n K_n$. Then proceeding by induction in n with the help of (3.54) we deduce that $\partial_n \partial_{n+1} = 0$ for each $n \geq 0$ since $\partial_0 \partial_1 = 0$ according to formulas (3.38) and (3.42).

There exists an opposite algebra A^{op}. The latter as an $\mathcal{T}$-linear space is the same, but with the multiplication $x \circ y = yx$ for each $x, y \in A^{op}$. Let $A^e := A \otimes_\mathcal{T} A^{op}$ denote the enveloping algebra of A. Apparently, $K_0 = A \otimes_\mathcal{T} A$ coincides with $A \otimes_\mathcal{T} A^{op}$ as a left and right A-module, hence the mapping $\partial_0 : K_0 \to K_{-1}$ provides the augmentation $\epsilon : A^e \to A$.

Thus identities (3.53) mean that the left complex $\mathcal{K}$

$$0 \leftarrow A \xleftarrow{\partial_0} K_0 \xleftarrow{\partial_1} K_1 \xleftarrow{\partial_2} K_2 \xleftarrow{\cdots} \xleftarrow{\partial_n} K_n \xleftarrow{\partial_{n+1}} K_{n+1} \leftarrow \cdots$$

is acyclic.

Example 3.3.
1. For the Cayley-Dickson algebra A_n over a field F of characteristic not equal to two let $G = G_n$ as the (multiplicative) metagroup consist of all elements bi_k with $b \in F_o$, $k = 0, 1, 2, \cdots$, where $i_0, i_1, i_2, \cdots$ are generators of the Cayley-Dickson algebra A_n, $2 \leq n \leq \infty$. Then $M = A_n^j$ is the module over $\mathbf{Z}[G]$, where $j \in \mathbf{N}$.

2. For a topological space U it is possible to consider the module $M = C(U, A_n^j)$ of all continuous mappings from U into A_n^j, $j \in \mathbf{N}$,

A_n^j is supplied with the box product topology.

3. If $(U, \mathcal{B}, \mu)$ is a measure space, where $\mu : \mathcal{B} \to [0, \infty)$ is a σ-additive measure on a σ-algebra $\mathcal{B}$ of a set U, for $\mathbf{F} = \mathbf{R}$ and $f_k = 1$ for each k, it is possible to consider the space $L_p((U, \mathcal{B}, \mu), A_n^j)$ of all L_p mappings from U into A_n^j, where A_n is taken relative to its norm induced by the scalar product $\operatorname{Re}(\bar{y}z) = (y, z), j \in \mathbf{N}, 1 \leq p \leq \infty$.

4. For an additive group H one can consider the trivial action of A on H. Therefore, the direct product $M \otimes H$ becomes an A-module for an A-module M. In particular, H may be a ring.

5. If there is another ring $\mathcal{S}$ and a homomorphism $\phi : \mathcal{S} \to \mathcal{T}$, then each left (or right) $\mathcal{T}$-module M can be considered as a left (or right correspondingly) $\mathcal{S}$-module by the rule: $bm = (\phi b)m$ (or $mb = m(\phi b)$ correspondingly) for each $b \in \mathcal{S}$ and $m \in M$.

Vice versa, if M is a right (or left) $\mathcal{S}$-module, then there exist the right (or left correspondingly) module $M_{(\phi)} = M \otimes_{\mathcal{S}} \mathcal{T}$ (or $_{(\phi)}M = \mathcal{T} \otimes_{\mathcal{S}} M$ correspondingly) called the right (or left correspondingly) covariant ϕ-extension of M. Similarly are defined the contravariant right and left extensions $M^{(\phi)} = \operatorname{Hom}_{\mathcal{S}}(\mathcal{T}, M)$ or $^{(\phi)}M$ for a right or left $\mathcal{S}$-module M respectively.

This also can be applied to a metagroup algebra $A = \mathcal{S}[G]$ over a commutative associative unital ring $\mathcal{S}$ as in Examples 3.1. Then changing a ring we get right $A_{(\phi)}$ or $A^{(\phi)}$ and left $_{(\phi)}A$ or $^{(\phi)}A$ algebras over $\mathcal{T}$. Then imposing the relation $ta = at$ for each $a \in A$ and $t \in \mathcal{T}$ provides a metagroup algebra over $\mathcal{T}$ which also has a two-sided $\mathcal{T}$-module structure. It will be denoted by $_{(\phi)}A_{(\phi)}$ or $^{(\phi)}A^{(\phi)}$ respectively. Particularly, this is applicable to cases when $\mathbf{Z}[F_\circ] \subset \mathcal{S}$ or ϕ is an embedding.

6. Let X_0 be a Hilbert space over $\mathbf{R}$ and let $\mathcal{A}_r$ be the Cayley-Dickson algebra over $\mathbf{R}$, where $r \geq 3$ (see Remark 2.2 and Examples 3.1). Then we put $X_k = i_k X_0$ for each k and $X = X_0 \oplus X_1 \oplus \cdots \oplus X_{2^r - 1}$, where $i_0 = 1$. Naturally, X can be supplied with a two-sided $\mathcal{A}_r$-module structure. Similarly can be considered a complexified Cayley-Dickson algebra $\mathcal{A}_{r,\mathbf{C}}$ and a Hilbert space X_0 over $\mathbf{C}$ and an $\mathcal{A}_{r,\mathbf{C}}$-two-sided module X, where $\mathcal{A}_{r,\mathbf{C}} = \mathcal{A}_r \oplus \mathbf{i}\mathcal{A}_r$ such that $\mathbf{i}^2 = -1$, $\mathbf{i}i_k = i_k\mathbf{i}$ for each $k = 0, \cdots, 2^r - 1$, where $\mathbf{C} = \mathbf{R} \oplus \mathbf{i}\mathbf{R}$ is the complex field. Hence X can be considered as the hypercomplex Hilbert space and the two-sided module with a scalar product $\langle x|y\rangle = (x_0, y_0) + \cdots + (x_{2^r-1}, y_{2^r-1})$ for each x and y in X such that a norm is $\|x\| = \sqrt{\langle x|x\rangle} \geq 0$, where $x = i_0 x_0 + \cdots + i_{2^r-1} x_{2^r-1}$, $x_j \in X_0$ for each j, where (x_j, y_j) denotes a scalar product on X_0.

For $r = \infty$ it can be taken l_2 direct sum of $\{X_k : k = 0, 1, 2, \cdots\}$ such that

$$X = l_2(\{X_k : k\}) = \left\{ z = (z_j \in X_j : j) : \|z\|^2 = \sum_{j=0}^{\infty} \|z_j\|^2 < \infty \right\}.$$

If take a metagroup $G = \{-i_j, i_j : 0 \leq j < 2^r, j \in \mathbf{Z}\}$, $3 \leq r \leq \infty$, then $X_j = i_j X_0$ and $X_j = -i_j X_0$, $i_1(i_2(i_4 X_0)) = (i_1 i_2)(i_4 X_0)$, $i_j X_j = X_0$, $(-i_j) X_j = X_0$, $i_2 X_1 = X_3$, etc., such that $X = \sum_{g \in G} Y_g$, where $Y_g = g X_0$ for each $g \in G$, because X_0 is the Hilbert space either over $\mathbf{R}$ or $\mathbf{C}$, $\mathcal{A}_r = \mathbf{R}[G]$ and $\mathcal{A}_{r,\mathbf{C}} = \mathbf{C}[G]$, where $\mathbf{R}$ is embedded into $\mathcal{A}_r$ as $i_0 \mathbf{R}$ and $\mathbf{C}$ is embedded into $\mathcal{A}_{r,\mathbf{C}}$ as $i_0 \mathbf{C}$.

Notation. Let $A = \mathcal{T}[G]$ be a metagroup algebra (see Definition 3.2). Put $L_0 = \mathcal{T}$, $L_1 = A$ and by induction $L_{n+1} = L_n \otimes_\mathcal{T} A$ for each natural number n.

If N is a two-sided A-module it can also be considered as a left A^e-module by the rule: $(x \otimes y)b := (xb)y$ for each $x \in A$, $y \in A^{op}$ and $b \in N$. $A^e = A \otimes_\mathcal{T} A^{op}$ is an enveloping algebra, where A^{op} denotes the opposite algebra of A.

Proposition 3.2. *If $\mathcal{K}$ is an acyclic left A-complex for a metagroup algebra $A = \mathcal{T}[G]$ as in Proposition 3.1 and M is a two-sided A-module satisfying conditions (3.27)-(3.30), then there exists a cochain complex* $\text{Hom}(\mathcal{L}, M)$:

$$0 \to \text{Hom}_\mathcal{T}(L_0, M) \xrightarrow{\epsilon^*} \text{Hom}_\mathcal{T}(L_1, M) \xrightarrow{\delta^1} \text{Hom}_\mathcal{T}(L_2, M) \xrightarrow{\delta^2}$$
$$\text{Hom}_\mathcal{T}(L_3, M) \xrightarrow{\delta^3} \text{Hom}_\mathcal{T}(L_4, M) \xrightarrow{\delta^4} \cdots \quad (3.55)$$

such that $f \in \text{Hom}_\mathcal{T}(L_1, M)$ is a cocycle if and only if f is a $\mathcal{T}$-linear derivation from A into M.

Proof. The notations of Example 3.2 and the Notation permit to write each generating element $(x_0, \cdots, x_{n+1})$ of K_n over $\mathcal{T}$ as

$$(x_0, \cdots, x_{n+1}) = t_{n+2}(x_0, \cdots, x_{n+1}; l(n+2), w(n+2)) \cdot ((x_0 \otimes (x_1, \cdots, x_n)) \otimes x_{n+1}) \quad (3.56)$$

$$(x_0, \cdots, x_{n+1}) = t_{n+2}(x_0, \cdots, x_{n+1}; l(n+2), w(n+2)) \cdot (z \otimes (x_1, \cdots, x_n)), \quad (3.57)$$

where $(x_1, \cdots, x_n)$ is a generating element in L_n for every $x_0, \cdots, x_{n+1}$ in G,

$$\{x_0, \cdots, x_{n+1}\}_{w(n+2)} = (x_0\{x_1, \cdots, x_n\}_{l(n)})x_{n+1},$$

where $z \in A^e$, $z = x_0 \otimes x_{n+1}^*$.

Each homomorphism $f \in \mathrm{Hom}_\mathcal{T}(L_n, M)$ is characterized by its values on elements $(x_1, \cdots, x_n)$, where $x_1,...,x_n$ belong to a metagroup G. Consider f as a $\mathcal{T}$-linear function from A^n into M. Since M satisfies conditions (3.27)-(3.30), then f has the decomposition

$$f(x_1, \cdots, x_n) = \sum_{g \in G} f_g(x_1, \cdots, x_n), \qquad (3.58)$$

where $f_g : G^n \to M_g$ for every g and $x_1, \cdots, x_n$ in G.

Therefore, the restrictions follow from conditions (3.27)-(3.30), which take into account the nonassociativity of G:

$$(xy) \cdot f_g(x_1, \cdots, x_n) = \mathsf{t}_3(x, y, g) \cdot (x \cdot (y \cdot f_g(x_1, \cdots, x_n))), \qquad (3.59)$$
$$\mathsf{t}_3(g, x, y) \cdot (f_g(x_1, \cdots, x_n) \cdot (xy)) = (f_g(x_1, \cdots, x_n) \cdot x) \cdot y, \qquad (3.60)$$
$$(x \cdot f_g(x_1, \cdots, x_n)) \cdot y = \mathsf{t}_3(x, g, y) \cdot (x \cdot (f_g(x_1, \cdots, x_n) \cdot y)), \qquad (3.61)$$

for every g and $x, y, x_1, \cdots, x_n$ in G, where coefficients t_3 are prescribed by formula (3.9), also

$$x \cdot f_g(x_1, \cdots, x_n) := x \cdot (f_g(x_1, \cdots, x_n)), \text{ and}$$
$$f_g(x_1, \cdots, x_n) \cdot y := (f_g(x_1, \cdots, x_n)) \cdot y. \qquad (3.62)$$

For $n = 0$ and $g = e$ naturally the identities are fulfilled:

$$(xy) \cdot f_e(\,) = x \cdot (y \cdot f_e(\,)), \ (f_e(\,) \cdot x) \cdot y = f_e(\,) \cdot (xy), \text{ and}$$
$$(x \cdot f_e(\,)) \cdot y = x \cdot (f_e(\,) \cdot y). \qquad (3.63)$$

There exists a coboundary operator taking into account the nonassociativity of the (multiplicative) metagroup G:

$$(\delta^n f)(x_1, \cdots, x_{n+1}) = \sum_{j=0}^{n+1} (-1)^j t_{n+1}(x_1, \cdots, x_{n+1}; l(n+1), u_{j+1}(n+1)) \cdot$$
$$[f, x_1, x_2, \cdots, x_{n+1}]_{j+1, n+1}, \qquad (3.64)$$

where

$$[f, x_1, \cdots, x_{n+1}]_{1,n+1} := x_1 \cdot f(x_2, \cdots, x_{n+1}),$$
$$\{x_1, \cdots, x_{n+1}\}_{u_1(n+1)} = x_1\{x_2, \cdots, x_{n+1}\}_{l(n)}; \qquad (3.65)$$
$$[f, x_1, \cdots, x_{n+1}]_{2,n+1} := f((x_1 x_2), \cdots, x_{n+1}),$$
$$\{x_1, \cdots, x_{n+1}\}_{u_2(n+1)} = \{(x_1 x_2), \cdots, x_{n+1}\}_{l(n)}; \cdots; \qquad (3.66)$$
$$[f, x_1, \cdots, x_{n+1}]_{n+1,n+1} := f(x_1, x_2, \cdots, (x_n x_{n+1})),$$
$$\{x_1, \cdots, x_{n+1}\}_{u_{n+1}(n+1)} = \{x_1, \cdots, (x_n x_{n+1})\}_{l(n)}; \qquad (3.67)$$
$$[f, x_1, \cdots, x_{n+1}]_{n+2,n+1} := f(x_1, x_2, \cdots, x_n) \cdot x_{n+1},$$
$$\{x_1, \cdots, x_{n+1}\}_{u_{n+2}(n+1)} = \{x_1, \cdots, x_{n+1}\}_{l(n+1)} = (\cdots((x_1 x_2)x_3)\cdots x_n)x_{n+1};$$
$$(3.68)$$

with $u_0(n+1) = l(n+1)$.

From G^{n+1} onto K_{n+1} the homomorphism $(\delta^n f)$ is extended by $\mathcal{T}$-linearity. On the other hand, condition (3.27) implies that: for each $b \in G$ there exists $h_{1,b}$ so that

$$h_{1,b} : K_{n+1} \to M_1 \text{ and } f_b = h_{1,b} L_b, \qquad (3.69)$$

where L_b is the left multiplication operator on b:

$$(h_{1,b} L_b)(x_1, \cdots, x_n) = b \cdot (h_{1,b}(x_1, \cdots, x_n)), \qquad (3.70)$$

for every $x_1, \cdots, x_n$ in G. Moreover, $zg = 0$ (or $gz = 0$) in $\mathbf{Z}[G]$ for $g \in G$ and $z \in \mathbf{Z}[G]$ if and only if $z = 0$ since G is a metagroup.

By virtue of Proposition 3.1, these formulas imply that $\delta^{n+1} \circ \delta^n = 0$ for each n since

$$(\delta^{n+1} \circ \delta^n f)(x_1, \cdots, x_{n+2}) = f(\partial_{n-1} \circ \partial_n(x_1, \cdots, x_{n+2})),$$

for every $x_1, \cdots, x_{n+2}$ in G. Thus the complex given by formula (3.55) is exact.

Particularly, $f \in \mathrm{Hom}_\mathcal{T}(L_0, M)$ is a cocycle if and only if

$$(\delta^0 f)(x) = xf(\ ) - f(\ )x = 0, \qquad (3.71)$$

for each $x \in G$. Notice that $\mathrm{Hom}_\mathcal{T}(L_0, M)$ is isomorphic with M.

One dimensional cochain $f \in \mathrm{Hom}_\mathcal{T}(L_1, M)$ is determined by a mapping $f : G \to M$. Taking into account formula (3.64) we infer that it is a cocycle if and only if

$$t_2(x, y; l(2), u_1(2)) \cdot x \cdot f(y) -$$
$$t_2(x, y; l(2), u_2(2)) \cdot f(xy) + t_2(x, y; l(2), u_3(2)) \cdot f(x) \cdot y$$
$$= x \cdot f(y) - f(xy) + f(x) \cdot y$$
$$= 0, \tag{3.72}$$

for each x and y in G. That is f is a derivation from the metagroup G into the G-module M. There is the embedding $\mathcal{T} \hookrightarrow A$ of $\mathcal{T}$ into A as $\mathcal{T}e$ since $e = 1 \in G$. Thus f has a $\mathcal{T}$-linear extension to a $\mathcal{T}$-linear derivation from A into M by the following formula:

$$f(xy) = x \cdot f(y) + f(x) \cdot y. \tag{3.73}$$

Remark 3.1. Suppose that conditions of Proposition 3.2 are fulfilled. A two-dimensional cochain is a 2-cocycle, if and only if

$$(\delta^2 f)(x_1, x_2, x_3) = \sum_{j=0}^{3} (-1)^j t_3(x_1, x_2, x_3; l(3), u_{j+1}(3)) \cdot [f, x_1, x_2, x_3]_{j+1,3}$$
$$= t_3(x_1, x_2, x_3; l(3), u_1(3)) \cdot x_1 \cdot f(x_2, x_3)$$
$$- t_3(x_1, x_2, x_3; l(3), u_2(3)) \cdot f((x_1 x_2), x_3)$$
$$+ t_3(x_1, x_2, x_3; l(3), u_3(3)) \cdot f(x_1, (x_2 x_3))$$
$$- t_3(x_1, x_2, x_3; l(3), u_4(3)) \cdot f(x_1, x_2) \cdot x_3$$
$$= 0,$$

that is

$$t_3(x_1, x_2, x_3) \cdot x_1 \cdot f(x_2, x_3) + t_3(x_1, x_2, x_3) \cdot f(x_1, (x_2 x_3))$$
$$= f((x_1 x_2), x_3) + f(x_1, x_2) \cdot x_3, \tag{3.74}$$

for each x_1, x_2 and x_3 in G.

As usually, $Z^n(A, M)$ denotes the set of all n-cocycles, the notation $B^n(A, M)$ is used for the set of n-coboundaries in $H_{\mathcal{T}}(L_n, M)$. Since as the additive group M is commutative, then there are defined groups of cohomologies $H^n(A, M) = Z^n(A, M)/B^n(A, M)$ as the quotient (additive) groups.

For $n = 0$, coboundaries are put zero and hence $H^0(A, M) \cong M^A$. In the case $n = 1$ a mapping $f : A \to M$ is a coboundary if there exists an element $m = h(\ ) \in M$ for which $f(x) = xm - mx$ for each $x \in A$. Such a derivation f is called an inner derivation of A defined by an element $m \in M$. The set of all inner derivations is denoted by $\text{Inn}_T(A, M)$.

From the cohomological point of view the additive group $H^1(A, M)$ is interpreted as the group of all outer derivations

$$H^1(A, M) \cong \text{Out}_T(A, M) \cong \text{Der}_T(A, M)/\text{Inn}_T(A, M),$$

where $Z^1(A, M) = \text{Der}_T(A, M)$; $\text{Inn}_T(A, M) = B^1(A, M)$, where the family of all derivations (T-homogeneous derivations) from X into a two-sided module M over T is denoted by $\text{Der}(X, M)$ (or $\text{Der}_T(X, M)$ respectively).

A two-cochain $f : G \times G \to M$ is a two-coboundary, if an one-cochain $h : G \to M$ exists such that for each x and y in G the identity is fulfilled:

$$f(x, y) = (\delta h)(x, y) = \sum_{j=0}^{2}(-1)^j t_2(x, y; l(2), u_{j+1}(2)) \cdot [h, x_1, x_2]_{j+1,2}$$
$$= x \cdot h(y) - h(xy) + h(x) \cdot y. \tag{3.75}$$

Let $A = T[G]$ be a metagroup algebra over a commutative associative unital ring T (see also Definitions 3.1 and 3.2).

Let M, N and P be left A-modules and a short exact sequence exists:

$$0 \to M \underset{\xi}{\to} P \underset{\eta}{\to} N \to 0, \tag{3.76}$$

where ξ is an embedding such that ξ and η are left A-homomorphisms. Then P is called an enlargement of a left A-module M with the help of a left A-module N. If there is another enlargement of M with the help of N:

$$0 \to M \underset{\xi'}{\to} P' \underset{\eta'}{\to} N \to 0, \tag{3.77}$$

such that an isomorphism $\pi : P \to P'$ exists for which $\pi\xi = \xi' 1_M$ and $1_N \eta = \eta' \pi$, then enlargements (3.76) and (3.77) are called equivalent, where $1_M : M \to M$ notates the identity mapping, $1_M(m) = m$ for each $m \in M$.

It is said that an enlargement clefts, if and only if there exists a left A-homomorphism $w : N \to P$ fulfilling the restriction $\eta w = 1_N$.

In the particular case, when $P = M \oplus N$, also ξ is an identifying mapping with the first direct summand and η is a projection on the second direct summand, an enlargement is called trivial.

Theorem 3.1. Let a nonassociative algebra A and left A-modules M and N be as in Remark 3.1. Then the family $T = \mathrm{Hom}_{\mathcal{T}}(N, M)$ can be supplied with a two-sided A-module structure such that $H^1(A, T)$ is the set of classes of modules M with the quotient module N. Moreover, $H^1(A, T)$ has the additive group structure.

Proof. The family $T = \mathrm{Hom}_{\mathcal{T}}(N, M)$ is a left module over a ring $\mathcal{T}$ and it can be supplied with a two-sided A-module structure: $\forall\, r \in T$ and $\forall\, n \in N$ and $\forall\, a \in A$:

$$(a \cdot r)(n) = a \cdot (r(n)) \text{ and } (r \cdot a)(n) = r(a \cdot n). \tag{3.78}$$

Notice that $H^1(A, T)$ has the additive group structure according to Remark 3.1.

By virtue of Proposition 3.1, each element $f \in Z^1(A, T)$ induces a (generalized) derivation by formula (3.73). Each zero dimensional cochain $m \in M$ provides an inner derivation $\delta^0 m(a) = am - ma$ due to formula (3.71). Then a one cocycle f induces an enlargement by formula (3.76) with $P = M \oplus N$ being the direct sum of left A-modules in which N is a submodule and with the left action of A on N: $a \circ n = a \cdot n + f(a) \cdot n$ for each $n \in N$ and $a \in A$. Suppose that a class of the one cocycle f is zero, that is an element $u \in T$ exists so that $f = \delta^1 u$. Then elements of the form $m + u(m)$ form its submodule M' isomorphic with M. Moreover, $P = N \oplus M'$ is the direct sum of A-modules. Thus an enlargement is trivial.

Vice versa, suppose that an enlargement given by formula (3.76) exists. That is a left A-homomorphism $\gamma : N \to P$ exists satisfying the restriction $\eta\gamma = 1_N$, where 1_N notates the identity mapping on N. It induces $f \in Z^1(A, T)$ such that $f(a)n = \gamma(a \cdot n) - (a \cdot \gamma)(n)$ for all $n \in N$ and for each element a of the algebra A.

Suppose that there is another enlargement which clefts, that is there exists a left $\mathcal{T}$-homomorphism $\omega : N \to P$ fulfilling the restriction $\eta\omega = 1_N$. We put $u(n) = \omega(n) - \gamma(n)$ for each $n \in N$, hence $u \in T$. Then $f_1 = \delta^1 u$ is a cocycle of zero class.

Theorem 3.2. Suppose that A is a nonassociative metagroup algebra over a commutative associative unital ring $\mathcal{T}$, a left A-module N and a two-sided A-module M are given (see Remark 3.1 and Theorem 3.1). Then $H^2(A, T)$ is the set of classes of enlargements of A with a kernel M such that $M^2 = \{0\}$ and with the quotient algebra A. Moreover, an action of A on M in this enlargement coincides with the structure of a two-sided A-module on M; $H^2(A, T)$ has the additive group structure.

Proof. If P is an enlargement with a kernel M such that $M^2 = 0$ and a quotient module $A = P/M$ and $a = p + M$ with $p \in P$, then $a \cdot m = p \cdot m$ and $m \cdot a = m \cdot p$ supply M with the two-sided A-module structure. Take a T-linear mapping $\gamma : A \to P$ inverse from the left to a natural epimorphism and put $f(a, b) = \gamma(ab) - \gamma(a)\gamma(b)$ for each a and b in A. Then we infer that
$$\gamma(a(bc)) = f(a, bc) + \gamma(a)\gamma(bc) = f(a, bc) + \gamma(a)(f(b, c) + \gamma(b)\gamma(c))$$
and
$$\gamma((ab)c) = f(ab, c) + \gamma(ab)\gamma(c) = f(ab, c) + (f(a, b) + \gamma(a)\gamma(b))\gamma(c),$$
consequently,
$$0 = \mathsf{t}_3(a, b, c)\gamma(a(bc)) - \gamma((ab)c)$$
$$= \mathsf{t}_3(a, b, c)f(a, bc) + \mathsf{t}_3(a, b, c)\gamma(a)(f(b, c) + \gamma(b)\gamma(c)) - f(ab, c)$$
$$- (f(a, b) + \gamma(a)\gamma(b))\gamma(c).$$

Taking into account that $\gamma(a)m = a \cdot m$ and $m\gamma(a) = m \cdot a$ for each $m \in M$ and $a \in A$, we deduce using formula (3.74) that
$$0 = \mathsf{t}_3(a, b, c) \cdot a \cdot f(b, c) - f(ab, c) + \mathsf{t}_3(a, b, c) \cdot f(a, bc) - f(a, b) \cdot c = (\delta^2 f)(a, b, c).$$

Thus $f \in B^2(A, M)$ and hence $f = (\delta^1 h)$ where $h \in C^1(A, M) := \mathrm{Hom}_T(A, M)$.

It remains to prove that the set S of all elements $\gamma(a) + h(a)$ forms in P a subalgebra isomorphic with A. From the construction of S it follows that S is a two-sided T-module. We verify that it is closed relative to the multiplication for all a and b in A:
$$(\gamma(a) + h(a))(\gamma(b) + h(b)) = \gamma(a)\gamma(b) + \gamma(a)h(b) + h(a)\gamma(b)$$
$$= \gamma(ab) - f(a, b) + ah(b) + h(a)b$$
$$= \gamma(ab) + h(ab) + ah(b) - h(ab) + h(a)b - f(a, b)$$
$$= \gamma(ab) + h(ab) + (\delta^1 h)(a, b) - f(a, b)$$
$$= \gamma(ab) + h(ab).$$

If there are given A, M, and f, then an enlargement P can be constructed as the direct sum $P = M \oplus A$ of two-sided T-modules and with the multiplication rule:
$$(m_1 + b_1)(m_2 + b_2) = m_1 b_2 + m_2 b_1 + f(b_1, b_2) + b_1 b_2,$$

for every m_1 and m_2 in M and b_1 and b_2 in A. It rest to verify that this multiplication rule is homogeneous over T and right and left distributive. At first, we evidently get that

$$(m_1 + b_1)(s(m_2 + b_2)) = (s(m_1 + b_1))(m_2 + b_2)$$
$$= s((m_1 + b_1)(m_2 + b_2))$$
$$= sm_1b_2 + sm_2b_1 + sf(b_1, b_2) + sb_1b_2,$$

and

$$(sp)(m_1 + b_1) = s(p(m_1 + b_1)),$$

for all $s, p \in T$; m_1, m_2 in M; and b_1, b_2 in A since $T \subset \mathcal{C}(A)$ and $f(s, p) = 0$. Moreover, we infer that

$$(m_1 + b_1)((m_2 + b_2) + (m_3 + b_3))$$
$$= (m_1 + b_1)((m_2 + m_3) + (b_2 + b_3))$$
$$= m_1(b_2 + b_3) + (m_2 + m_3)b_1 + f(b_1, b_2 + b_3) + b_1(b_2 + b_3)$$
$$= m_1b_2 + m_1b_3 + m_2b_1 + m_3b_1 + f(b_1, b_2) + f(b_1, b_3) + b_1b_2 + b_1b_3$$
$$= (m_1 + b_1)(m_2 + b_2) + (m_1 + b_1)(m_3 + b_3),$$

and analogously

$$((m_1+b_1)+(m_2+b_2))(m_3+b_3) = (m_1+b_1)(m_3+b_3)+(m_2+b_2)(m_3+b_3),$$

for all m_1, m_2, m_3 in M; and b_1, b_2, b_3 in A.

In the proof of Theorem 3.1, it was shown that $T = \text{Hom}_T(N, M)$ is the two-sided A-module. On the other hand, by Remark 3.1, $H^2(A, T)$ has the additive group structure.

Definition 3.4. Let M, P, and N be two-sided A-modules, where A is a nonassociative metagroup algebra over a commutative associative unital ring T. An A-homomorphism (isomorphism) $f : M \to P$ is called a right (operator) A-homomorphism (isomorphism) if it is such for M and N as right A-modules, that is

$$f(x + y) = f(x) + f(y), \text{ and}$$
$$f(xa) = f(x)a,$$

for each x, y in M and $a \in A$ (see also Definition 3.3). If an algebra A is specified it may be written shortly a homomorphism (isomorphism) instead of an A-homomorphism (an A-isomorphism respectively).

An enlargement (P, η) of M by N is called right inessential, if a right isomorphism $\gamma : N \to P$ exists satisfying the restriction $\eta\gamma|_N = 1|_N$.

Theorem 3.3. Suppose that M is a two-sided A-module, where A is a nonassociative metagroup algebra over a commutative associative unital ring $\mathcal{T}$. Then for each $n \geq 0$ there exists a two-sided A-module P_n such that $H^{n+1}(A, M)$ is isomorphic with the additive group of equivalence classes of right inessential enlargements of M by P_n.

Proof. Consider two right inessential enlargements (E_1, η_1) and (E_2, η_2) of M by N, where ξ_1 and ξ_2 are embeddings of M into E_1 and E_2 correspondingly. Take a submodule Q of $E_1 \oplus E_2$ consisting of all elements (x_1, x_2) satisfying the condition: $\eta_1(x_1) = \eta_2(x_2)$. Then a quotient module Q/T exists, where $T = \{(\xi m, -\xi m) : m \in M\}$. Therefore $(\xi_1 M \oplus \xi_2 M)/T$ is isomorphic with M and homomorphisms η_1 and η_2 induce a homomorphism η of Q/T onto N. Hence the submodule $\ker(\eta)$ is isomorphic with M. Then an addition of enlargements is prescribed by the formula:

$$(E_1, \eta_1) + (E_2, \eta_2) := (Q/T, \eta).$$

Evidently sums of equivalent enlargements are equivalent.

For an enlargement (E, η) of M by N one takes the direct sum of modules $E \oplus M$ and puts T_b to be its submodule consisting of all elements $(\xi m, -b\xi m)$ with $m \in M$, where ξ is an embedding of M into E, $b \in \mathcal{T}$. Therefore, a homomorphism η induces a homomorphism $_b\eta$ of $(E \oplus M)/T_b$ onto N since the mapping $(\xi m, m) \mapsto b\xi m + m$ is a homomorphism of $(\xi M) \oplus M$ onto M, also the ring $\mathcal{T}$ is commutative and associative. This induces an enlargement of M by N denoted by $(_bE, _b\eta)$ and hence an operation of scalar multiplication of an enlargement (E, η) on $b \in \mathcal{T}$. From this construction, it follows that equivalent enlargements have equivalent scalar multiplies on $b \in \mathcal{T}$.

Let P_n be a $\mathcal{T}$-linear span of all elements $(x_1, \cdots, x_{n+1})$ with $x_1, \cdots, x_{n+1}$ in G such that

$$((bx_1), x_2, \cdots, x_{n+1}) = (x_1, \cdots, (bx_{n+1})),$$

for each $b \in \mathcal{T}$. Next, we put

$$(x_1, \cdots, x_{n+1}) \cdot y := t_{n+2}(x_1, \cdots, x_{n+1}, y; l(n+2), u_{n+2}(n+2)) \cdot$$
$$(x_1, \cdots, x_n, (x_{n+1}y)), \qquad (3.79)$$

and

$$y \cdot (x_1, \cdots, x_{n+1}) = \sum_{j=1}^{n+1} (-1)^{j+1} \cdot t_{n+2}(y, x_1, \cdots, x_{n+1}; u_1(n+2), u_{j+1}(n+2)) \cdot$$
$$\langle y, x_1, x_2, \cdots, x_{n+1} \rangle_{j,n+2}, \quad (3.80)$$

(see also Notation (3.39)-(3.41) and (3.65)-(3.68)) for every $y, x_1, \cdots, x_{n+1}$ in G. That is P_n is the two-sided A-module, where A has the unit element.

We denote by $R_n = R(P_n, M)$ the family of all right homomorphisms of P_n into M. For each $p \in R_n$ let an arbitrary element $\dot{p} \in C^n(A, M)$ in the additive group of all n cochains (that is, n times $\mathcal{T}$-linear mappings of A into M) on A with values in M be prescribed by the formula

$$\dot{p}(a_1, \cdots, a_n) = p(a_1, \cdots, a_n, 1),$$

for all $a_1, \cdots, a_n$ in A. Consequently,

$$(\dot{p}(a_1, \cdots, a_n)) \cdot y = p(a_1, \cdots, a_n, y),$$

for each $y \in A$ since $t_{n+3}(x_1, \cdots, x_{n+1}, 1, g; l(n+3), u_{n+3}(n+3)) = 1$ for all $x_1, \cdots, x_{n+1}$ and g in G. This makes the mapping $p \mapsto \dot{p}$ an $\mathcal{T}$-linear isomorphism of R_n onto $C^n(A, M)$.

Supply $C^n(A, M)$ with a two-sided A-module structure:

$$(x_0 \cdot f)(x_1, \cdots, x_n) = x_0 \cdot (f(x_1, \cdots, x_n)), \quad (3.81)$$

and

$$(f \cdot x_0)(x_1, .., x_n) = \sum_{k=0}^{n-1} (-1)^k t_{n+1}(x_0, x_1, \cdots, x_n; u_1(n+1), u_{k+2}(n+1)) \cdot$$
$$f(x_0, \cdots, x_k x_{k+1}, \cdots, x_n) + (-1)^n (f(x_0, \cdots, x_{n-1})) \cdot x_n, \quad (3.82)$$

for each $f \in C^n(A, M)$ and all $x_0, x_1, \cdots, x_n$ in G extending f by $\mathcal{T}$-linearity on A from G, where $u_j(n+1)$ are given by formulas (3.65)-(3.68). Thus the mapping $p \mapsto \dot{p}$ is an operator isomorphism, consequently, $H^p(A, R_n)$ is isomorphic with $H^p(A, C^n(A, M))$ for each integers n and p such that $n \geq 0$ and $p \geq 0$. On the other hand, $H^p(A, C^n(A, M))$ is isomorphic with $H^{p+n}(A, M)$ for each $p \geq 1$, hence $H^p(A, R_n)$ is isomorphic with $H^{p+n}(A, M)$.

By virtue of Theorem 3.1 applied with $p = 1$ we infer that $H^{n+1}(A, M)$ is isomorphic with the additive group of equivalence classes of right inessential enlargements of M by P_n.

Derivations of Operator Algebras on Hypercomplex Hilbert Spaces ... 99

Theorem 3.4. Let M be a two-sided A-module, where A is a nonassociative metagroup algebra over a commutative associative unital ring T. Then to each $(n+1)$-cocycle $f \in Z^{n+1}(A, M)$ an enlargement of M by a two-sided A-module P_n corresponds such that f becomes a coboundary in it.

Proof. An $(n+1)$-cocycle $f \in Z^{n+1}(A, M)$ induces an enlargement (E, η) of M by P_n due to Theorem 3.3. An element h in $Z^1(A, R_n)$ corresponding to f is characterized by the equality:

$$(h(x_1))(x_2, \cdots, x_{n+1}, 1) = t_{n+1}(x_1, \cdots, x_{n+1}; u_1(n+1), l(n+1)) \cdot f(x_1, \cdots, x_{n+1}),$$

for all $x_1, \cdots, x_{n+1}$ in G. This enlargement (E, η) as the two-sided A-module is $P_n \oplus M$ such that

$$x_1 \cdot ((x_2, \cdots, x_{n+1}, 1), 0) = (x_1 \cdot (x_2, \cdots, x_{n+1}), f(x_1, \cdots, x_{n+1})).$$

Let $\gamma(a_1, \cdots, a_n) = (a_1, \cdots, a_n, 0)$ for all $a_1, \cdots, a_n$ in A. Therefore we deduce that

$$f(x_1, \cdots, x_{n+1}) = t_{n+1}(x_1, \cdots, x_{n+1}; l(n+1), u_1(n+1)) \cdot$$
$$\{x_1 \cdot \gamma(x_2, \cdots, x_{n+1}, 1) - \gamma(x_1 \cdot (x_2, \cdots, x_{n+1}, 1))\}.$$

There exists a n-cochain $v \in C^n(A, E)$ defined by $v(a_1, \cdots, a_n) = ((a_1, \cdots, a_n, 1), 0)$ for all $a_1, \cdots, a_n$ in A. Thus $f = \delta v$.

Theorem 3.5. Let A be a nonassociative metagroup algebra over a commutative associative unital ring T. Then an algebra B over T exists such that B contains A and each T-homogeneous derivation $d: A \to A$ is the restriction of an inner derivation of B.

Proof. Naturally, an algebra A has the structure of a two-sided A-module. In view of Proposition 3.2, each derivation of the two-sided algebra A can be considered as an element of $Z^1(A, A)$.

Applying Theorem 3.4 by induction one obtains a two-sided A-module Q containing M for which an arbitrary element of $Z^{n+1}(A, M)$ is represented as the coboundary of an element of $C^n(A, Q)$. At the same time M and Q satisfy conditions (3.27)-(3.30). This implies that the natural injection of $H^{n+1}(A, M)$ into $H^{n+1}(A, Q)$ maps $H^{n+1}(A, M)$ into zero.

Therefore, a two-sided A-module E exists which as a two-sided $\mathcal{T}$-module is a direct sum $A \oplus P$ and P is such that for each $f \in Z^1(A, A)$ there exists an element $p \in P$ generally depending on f with the property $f(a) = a \cdot p - p \cdot a$. To the algebra A the metagroup G corresponds. Enlarging P if necessary we can consider that to P a metagroup G also corresponds in such a manner that properties (3.27)-(3.30) are fulfilled.

Now we take $A \oplus P$ as the underlying two-sided $\mathcal{T}$-module of B and supply it with the multiplication $(a_1, p_1)(a_2, p_2) := (a_1 a_2, a_1 \cdot p_2 + p_1 \cdot a_2)$ as the semidirect product for each a_1, a_2 in A and p_1, p_2 in P. An embedding ξ of A into B is $\xi(a) = (a, 0)$ for each a in A. This implies that

$$f(a) = (a, 0)(0, p) - (0, p)(a, 0) = a(0, p) - (0, p)a.$$

Theorem 3.6. Suppose that A is a nonassociative metagroup algebra of finite order over a commutative associative unital ring $\mathcal{T}$ and M is a finitely generated two-sided A-module. Then M is semisimple if and only if its cohomology group is null $H^n(A, M) = 0$ for each natural number $n \geq 1$.

Proof. Certainly, if E is an A-module and N its A-submodule, then a natural quotient morphism $\pi : E \to E/N$ exists. Therefore, an enlargement (E, η) of a two-sided A-module M by a two-sided A-module N is inessential if and only if there is a submodule T in E complemented to $\xi(M)$ such that T is isomorphic with $E/\xi(M)$, where ξ is an embedding of M into E. If M is semisimple, then it is either simple or a finite product of simple modules since M is finitely generated. For a finitely generated module E and its submodule N the quotient module E/N is not isomorphic with E since the algebra A is of finite order over the commutative associative unital ring $\mathcal{T}$.

By virtue of Theorems 3.3 and 3.4 if for an algebra A its corresponding finitely generated two-sided A-modules are semisimple, then its cohomology groups of dimension $n \geq 1$ are zero.

Vise versa, suppose that $H^n(A, M) = 0$ for each natural number $n \geq 1$. Consider a finitely generated two-sided A-module E and its two-sided A-submodule N. At first, we take into account the right A-module structure E_r of E with the same right transformations, but with zero left transformations. Then the left inessential (E_r, η_r) enlargement of M_r by $N_r = E_r/\xi_r(M_r)$ exists, where $\eta_r : E_r \to E_r/\xi_r(M_r)$ is the quotient mapping and ξ_r is an embedding of M_r into E_r. From Theorem 3.3, it follows that the enlargement (E_r, η_r) is right inessential. Analogously considering left A-module structures E_l and M_l we infer that (E_l, η_l) is also left inessential.

Remark 3.2. Let A be a nonassociative metagroup algebra over a commutative associative unital ring $\mathcal{T}$ of characteristic $\mathrm{char}(\mathcal{T})$ other than two and three. There exists its opposite algebra A^{op}. The latter as an F-linear space is the same, but with the multiplication $x \circ y = yx$ for each $x, y \in A^{op}$. To each element $h \in A$ or $y \in A^{op}$ there is posed a left multiplication operator L_h by the formula $xL_h = hx$ or a right multiplication operator $xR_y = xy$ for each $x \in A$ respectively. Having the anti-isomorphism operator

$$S : A \to A^{op}, A \ni x \mapsto xS \in A^{op}, S(xy) = S(y)S(x),$$

we get

$$R_h S = SL_{hS} \text{ and } L_h S = SR_{hS}, \tag{3.83}$$

for each x and h in A. Then taking into account (3.83) analogously to formula (3.82), we put

$$x_0 \cdot (L_{x_1}, R_{x_2}) = \mathsf{t}_3^{-1}(x_0, x_1, x_2) \cdot (x_0 L_{x_1}) L_{x_2} S - x_0 (L_{x_1} R_{x_2}) + \mathsf{t}_3^{-1}(x_0, x_1, x_2) \cdot (x_0 R_{x_1} S) R_{x_2}$$

Then taking into account multipliers t_3 this gives

$$x_0 \cdot (L_{x_1}, R_{x_2}) = x_0 (L_{x_1} L_{x_2} S - L_{x_1} R_{x_2} + R_{x_1} S R_{x_2}), \tag{3.84}$$

for all x_0, x_1, x_2 in G. Next symmetrically $S(x_0 \cdot (L_{x_1}, R_{x_2}))$ provides the formula for $(L_{y_1}, R_{y_2}) \cdot y_0$ for each y_0, y_1 and y_2 in G. We consider the enveloping algebra $A^e = A \otimes_{\mathcal{T}} A^{op}$. Extending these rules by $\mathcal{T}$-linearity on A and $A \otimes_{\mathcal{T}} A^{op}$ from G one supplies the tensor product $M = A^e$ over $\mathcal{T}$ with the two-sided A-module structure.

Corollary 3.1. Let A be a semisimple nonassociative metagroup algebra of finite order over a commutative associative unital ring $\mathcal{T}$ of characteristic $\mathrm{char}(\mathcal{T})$ other than two and three and let M be a two-sided A-module described in Remark 3.2. Then $H^n(A, M) = 0$ for each natural number $n \geq 1$.

Proof. Since A is semisimple, then the module M from Remark 3.2 is semisimple, consequently, from Theorem 3.6 the statement of this corollary follows.

Corollary 3.2. Let A be a semisimple nonassociative metagroup algebra of finite order over a commutative associative unital ring $\mathcal{T}$ of characteristic $\mathrm{char}(\mathcal{T})$ other than two and three and let M be a two-sided A-module described in Remark 3.2. Then $H^n(A, M) = 0$ for each natural number $n \geq 1$.

Proof. Since A is semisimple, then the module M from Remark 3.2 is semisimple, consequently, from Theorem 3.6 the statement of this corollary follows.

Example 3.4. Let a commutative associative unital ring T be of characteristic other than 2 and 3. Consider a Cayley-Dickson algebra A_n over the ring T with $A_0 = T$ and $n \geq 3$. Then one can take its basis $\{i_0, i_1, \cdots, i_{2^n-1}\}$ as the two-sided T module composed of doubling generators i_{2^r} for each $1 \leq r < n$ with $i_0 = 1$ and their ordered products: $i_1 i_2 = i_3$, $i_k i_{2^r} = i_{k+2^r}$ for each $0 \leq k < 2^r$, etc. Suppose that $i_k i_j = f_{k,j} i_j i_k$ with $f_{k,j}$ belonging a multiplicative group $F_\circ$ for each j and k such that $F_\circ \subset T$. Then, we put $G_n = \{i_0, i_1, \cdots, i_{2^n-1}\} \times F_\circ$, consequently, $T[G_n]$ is isomorphic with A_n. Then in view of Corollary 3.2 for each $n \geq 3$ and the Cayley-Dickson algebra A_n there exist a metagroup G_n and a module M_n such that $T[G_n] = A_n$ and $\mathrm{Der}_T(A_n) = B^1(A_n, M_n)$.

Mention that the minimal subalgebra $\mathrm{alg}_T(i_k, i_l, i_j)$ over T generated by any three generators i_k, i_l, i_j is alternative, where k, l and j belong to the set $\{0, 1, \cdots, 2^n - 1\}$. On the other hand, $xL_h = \sum_{j,k} x_j h_k (i_j L_{i_k})$ and $xR_h = \sum_{j,k} x_j h_k (i_j R_{i_k})$, where x an h are in A_n, $x = x_0 i_0 + \cdots + x_{2^n-1} i_{2^n-1}$, $x_0 \in T, \cdots, x_{2^n-1} \in T$. Evidently, if θ is an automorphism of A and d is a derivation from A into $A \times A^{op}$, then they induce new derivation $(\theta \times \theta^{op}) d\theta^{-1}$.

For $n \geq 3$, Corollary 3.2 and formula (3.84) provide all derivations $d = [L_x, L_z] + [L_x, R_z] + [R_x, R_z]$ from A into $M = A \times A^{op}$, where x and z are in A since in this case $Sz = \bar{z}$ and T is the center of A_n, $[x, z] = [\mathrm{Im}(x), \mathrm{Im}(z)]$, where $\mathrm{Im}(x) = (x - \bar{x})/2$, $[x, z] = xz - zx$. Thus all derivations of A_n are inner.

Mention that for $n \geq 4$, there are relations caused by divisors of zero, so the group $\mathrm{Der}_T(A_{n+1})$ is not very abundant in comparison with $\mathrm{Der}_T(A_n)$. Bases in Lie algebras of derivations of Cayley-Dickson algebras were described in details in [5, 29, 30]. In [30] the results are more general than in [5, 29] where derivations were considered using canonical embeddings of A_n into A_{n+1}, but there may be different embeddings of A_n into A_{n+1} and different bases of generators can be chosen up to automorphisms of algebras. In particular, for $n = 2$ the quaternion algebra A_2 is associative and expressions for derivations simplify since $[L_x, R_z] = 0$ and hence $d = \mathrm{ad}(s)$ in this case, where $s = [x, z]$.

Derivations of Operator Algebras on Hypercomplex Hilbert Spaces ... 103

Theorem 3.7. Let A_n be a family of simple nonassociative metagroup algebras of finite order $(A_n : \mathcal{T})$ over a commutative associative unital ring $\mathcal{T}$ of $\text{char}(\mathcal{T}) \neq 2$, where Λ is a directed set, $\text{card}(\Lambda) = \aleph_0$, $h_n^m : A_m \hookrightarrow A_n$ is an algebraic $\mathcal{T}$-linear embedding for each $m < n$ in Λ. Let also an algebra A of infinite order over $\mathcal{T}$ be a limit $A = \lim\{A_m, h_n^m, \Lambda\}$ of a direct homomorphism system $\{A_m, h_n^m, \Lambda\}$ such that for each $m \in \Lambda$ a set $\Upsilon(m) := \{n : n \in \Lambda, m < n\}$ is infinite, $h_k^n(A_n) \neq A_k$ for each $n < k$ in $\Upsilon(m)$ and $\text{card}(\Upsilon(m)) = \aleph_0$ for each $m \in \Lambda$. Then $H^1(A, A \times A^{op}) \neq 0$.

Proof. Without loss of generality, we can consider $h_n^m(A_m) \neq A_n$ for each $m < n$ in Λ since $(A : \mathcal{T}) = \infty$ and $(A_n : \mathcal{T}) < \infty$ for each $n \in \Lambda$. Therefore, for each $l > 1$ and each $m \in \Lambda$ there exists $n \in \Lambda$ such that $m < n$ and with orders $(A_n : \mathcal{T}) \geq l(A_m : \mathcal{T})$. Put

$$C_n = \{y \in Y : \forall a \in h^n(A_n) \; \exists p \in h^n(F_{\circ,n}) \; ay = pya\},$$

for each $n \in \Lambda$, where $Y = \hat{A} \otimes \check{A}$ is the two-sided A-module with the first multiplier $\hat{A}$ being the algebra A considered as the left A-module and the second multiplier $\check{A}$ being the algebra A considered as the right A-module since $F_{\circ,n} \subset \mathcal{T}$ for each $n \in \Lambda$. Moreover, we have $Y = \lim\{\hat{A}_m \otimes \check{A}_m, s_n^m, \Lambda\}$, where $s_n^m = h_n^m \times h_n^m$ for each $m < n$ in Λ, $\hat{A}_m$ (or $\check{A}_m$) denotes the algebra A_m considered as the left (or right correspondingly) A_m-module. Evidently, $\mathcal{T} \otimes \mathcal{T} \subset C_n$ for each $n \in \Lambda$ since $1 \in A_n$, while $\mathcal{T}^{op}$ is isomorphic with $\mathcal{T}$ for the commutative associative unital ring $\mathcal{T}$ and $\mathcal{T} \otimes \mathcal{T}^{op} \subset A_n \otimes A_n^{op}$ for each n. From the definition of the set C_n and the embedding existence $h_n^m : A_m \hookrightarrow A_n$ for each $m < n \in \Lambda$ it follows that $C_n \subset h_n^m C_m$.

There is an embedding $h^m : A_m \hookrightarrow A$ for each $m \in \Lambda$. Then A_m has left and right complements $U_{m,n}$ and $V_{m,n}$ in A_n for each $m < n$ in Λ. Therefore,

$$Y_n = s_n^m(Y_m) \oplus (h_n^m(\hat{A}_m) \otimes V_{m,n}) \oplus (U_{m,n} \otimes h_n^m(\check{A}_m)) \oplus (U_{m,n} \otimes V_{m,n}),$$

as two-sided A_m-modules for each $m < n$ in Λ. Take $Q_{m,n} = C_m \cap (U_{m,n} \otimes V_{m,n})$ and the $\mathcal{T}$-linear projection $P_{m,n}$ of $C_n \cap Y_n$ into $U_{m,n} \otimes V_{m,n}$ for each $m < n$ in Λ.

To each $x \in A$ (or $z \in A^{op}$) we can pose the left L_x (or right R_z correspondingly) multiplication operator, where $yL_xR_z = (xy)z$. We have that $F_{\circ,n}$ is a normal subgroup of G_n for each $n \in \Lambda$, where $F_{\circ,n}$ is embedded into G_n as $F_{\circ,n}e_n$, where e_n denotes the neutral element of the multiplicative metagroup G_n. Since G_m is a metagroup and h_n^m is an algebraic embedding, then $h_n^m(G_m)$ is also a metagroup for each $m < n$ in Λ. This induces an embedding $h_n^m : G_m \hookrightarrow G_n$ for each $m < n$ in Λ. Hence the multiplicative group $F_\circ = \lim\{F_{\circ,m}, h_n^m, \Lambda\}$ is the normal subgroup in a metagroup G, where $G = \lim\{G_m, h_n^m, \Lambda\}$. Hence there exists the quotient mapping $\tau : G \to P$ with $G/F_\circ =: P$ and P is associative, that is a group. From the condition $(ab)c = \mathsf{t}_3(a,b,c)a(bc)$ for each a, b and c in G_n for each $n \in \Lambda$, where a function t_3 has values in $F_{\circ,n}$ and it is for a metagroup G_n, we deduce that C_m contains the $\mathcal{T}$-linear span of the operator set $\mathcal{E} := \{(a,b) \in G \times G^{op} : \tau(ab) = \tau(ba)\}$.

From the definition of a metagroup G_n it follows that the quotient metagroup $J_n := G_n/F_{\circ n}$ is associative for each $n \in \Lambda$. That is J_n is a group. Moreover, J_m has embedding as a proper subgroup into J_n for each $m < n$ in Λ. This implies that for each $m \in \Lambda$ there exists $n = n(m)$ such that $m < n$ in Λ and an element $x_{m,n}$ in $Q_{m,n} \setminus P_{m,n}$ exists. Therefore, $x_{m,n} \in Q_{m,n} \subset Y_n \cap C_m$ and $x_{m,n} \notin Y_n + C_n$.

Since the set Λ is directed, then for each $m_1, \cdots, m_k$ there exists $n \in \Lambda$ so that $m_j < n$ for each $j = 1, \cdots, k$.

There exists a topology on Y a base at zero of which is composed of sets C_m, where $m \in \Lambda$. Let $m \in \Lambda$ be some marked elements and let $n_1 = n(m)$, $n_2 = n(n_1), \cdots, n_{k+1} = n(n_k)$ for each $k \in \mathbf{N}$. Suppose that a series $\sum_{k=1}^\infty z_k$ converges to an element z in Y, where $z_k = x_{m,n_k}$ for each $k \in \mathbf{N}$. From the preceding construction the inclusions $-\sum_{k=b+1}^\infty z_k \in C_{n_{j+1}}$ and $\sum_{k=1}^{j-1} z_k \in Y_{n_j}$ and $\sum_{k=j+1}^b z_k \in C_{n_{j+1}}$ follow for each $b > j$ since $x_{m,n} \in Q_{m,n} \subset Y_n \cap C_m$. This implies that $z_j \in C_{n_{j+1}} + Y_{n_j} - z$. Particularly, for j satisfying the inclusion $z \in Y_{n_j}$ one gets $z_j \in C_{n_{j+1}} + Y_{n_j}$, but this contradicts the property $x_{m,n} \notin Y_n + C_n$. Thus Y is not complete. But then $dx := \sum_{k=1}^\infty (z_k x - x z_k)$ is a non-inner derivation of A into Y. This mapping d is correctly defined and nonnull since $z_k = x_{m,n_k}$ for each $k \in \mathbf{N}$ and $x_{m,n} \in Q_{m,n} \setminus P_{m,n}$ for each $m < n = n(m)$ in Λ. Indeed, for each $x \in A$ there exists $n \in \Lambda$ such that $x \in h^n(A_n)$. At the same time $z_k \in C_{n_k}$ for each k and $C_{n_j} \subset C_{n_k}$ for each $k < j$ in Λ. Therefore a natural number l exists such that $dx = \sum_{k=1}^l (z_k x - x z_k)$. If $dx = mx - xm$ for some fixed $m \in Y$, then $m - \sum_{k=1}^b z_k \in C_{n_b}$ for each $b \in \mathbf{N}$ since $mx - xm = \sum_{k=1}^b (z_k x - x z_k)$ for each $x \in h^b(A_b)$.

Theorem 3.8. Let A be a locally finite semisimple nonassociative metagroup algebra over a commutative associative unital ring $\mathcal{T}$ of characteristic $char(\mathcal{T}) \neq 2$ and let its order over $\mathcal{T}$ be $(A : \mathcal{T}) = \aleph_0$. Then $H^1(A, A \otimes A^{op}) \neq$

0.

Proof. If an algebra A is not simple, then it has the decomposition into the direct sum $A = \bigoplus_k A_k$ of simple algebras A_k. If $(A_k : \mathcal{T}) < \aleph_0$ for each k, then according to the preceding Theorem 3.7, $H^1(A, A \otimes A^{op}) \neq 0$ since there are embeddings $h_n^m : \bigoplus_{k=1}^m A_k \hookrightarrow \bigoplus_{k=1}^n A_k$ for each $m < n$ and taking $\Lambda = \mathbf{N}$.

Thus it is sufficient to consider the case when A is simple and $(A : \mathcal{T}) = \aleph_0$ since the remaining case is such that there exists k for which $(A_k : \mathcal{T}) = \aleph_0$. By the definition of a locally finite algebra for each finite set $x_1, \cdots, x_n$ of elements in A there exists a subalgebra $\mathrm{alg}_\mathcal{T}(x_1, \cdots, x_n)$ of finite order over $\mathcal{T}$ generated by these elements, $(\mathrm{alg}_\mathcal{T}(x_1, \cdots, x_n) : \mathcal{T}) < \aleph_0$ and $\mathrm{alg}_\mathcal{T}(x_1, \cdots, x_n) \subset A$. Therefore there exists a direct homomorphism system of simple subalgebras over the ring $\mathcal{T}$ satisfying conditions of Theorem 3.7 with $\mathrm{card}(\Lambda) = \aleph_0$ and $\mathrm{card}(\Upsilon_m) = \aleph_0$ for each $m \in \Lambda$ since $(A : \mathcal{T}) = \aleph_0$. By virtue of Theorem 3.7, $H^1(A, A \otimes A^{op}) \neq 0$.

Example 3.5. If A_∞ is a Cayley-Dickson algebra of infinite order over a commutative associative unital topological ring $\mathcal{T}$ of $\mathrm{char}(\mathcal{T}) \neq 2$, also $A_0 = \mathcal{T}$, then $A = \lim\{A_m, h_n^m, \mathbf{N}\}$. Each A_n is of finite order over $\mathcal{T}$ and inherits a topology from $\mathcal{T}^{2^n}$. In this example, the Cayley-Dickson algebra A is supplied with the direct limit topology. Therefore, Theorems 3.7 and 3.8 are applicable in this case, hence $H^1(A, A \otimes A^{op}) \neq 0$.

The results below also treat infinite dimensional Cayley-Dickson algebras over fields supplied with norm topologies.

Theorem 3.9. Let A be an nonassociative metagroup algebra over a commutative associative unital ring $\mathcal{T}$ with a metagroup G such that $G \cap \mathcal{T} = F_\circ$, and let H be a proper normal submetagroup isomorphic with G such that the quotient metagroup G/H is infinite and $F_\circ \subset H$. Then outer $\mathcal{T}$-linear automorphisms of A exist and the cardinality of their family is not less than 2^ξ, where $\xi = \mathrm{card}(G/F_\circ)$.

Proof. By the conditions of this theorem G/H is infinite, that is the cardinality $\mathrm{card}(G/H) \geq \aleph_0$. Moreover, we infer that $G/H = (G/F_\circ)/(H/F_\circ)$, since $F_\circ \subset H \subset G$ and H is isomorphic with G. At the same time the quotients $J := G/F_\circ$ and $J_\infty := H/F_\circ$ are groups. Thus G/H is the group.

If an element b of a group P is infinite divisible, that is for each $n \in \mathbf{N}$ there exists $a \in P$ having the property $a^n = b$, we put $\mathrm{ord}(b) = 0$. Then

$ord(b) = m > 0$ is a least natural number m such that $b^m = e$ is a neutral element. Otherwise $ord(b) = \infty$, when b is neither infinite divisible nor is of finite order. From the theorem about transfinite induction [41], it follows that there are sets Υ and Υ_∞ such that

$$\Upsilon \subset J \text{ and } \Upsilon_\infty \subset J_\infty, \tag{3.85}$$

satisfying the properties: $card(\Upsilon) \geq \aleph_0$ and $card(\Upsilon_\infty) \geq \aleph_0$, $card(\Upsilon) = card(\Upsilon_\infty)$, also

$$card(J \setminus gr_J(\Upsilon)) \leq card(J); \tag{3.86}$$

$$\Upsilon = \bigcup_{b \in \Psi} b\Upsilon_\infty, \tag{3.87}$$

where Ψ is a set of the cardinality $card(\Psi) \geq \aleph_0$, $b\Upsilon_\infty \cap c\Upsilon_\infty = \emptyset$ for each $b \neq c$ in Ψ; $\forall b \in \Upsilon$ $b \notin gr_J(\Upsilon \setminus \{b\})$ and $\forall b \in \Upsilon_\infty$ $b \notin gr_{J_\infty}(\Upsilon_\infty \setminus \{b\})$ since $card(B \times B) = card(B)$ for $card(B) \geq \aleph_0$, where $gr_J(K)$ denotes an intersection of all subgroups in J containing a subset K, where $K \subset J$. The cardinality of the family of inner automorphisms of G is not greater than ξ since each inner automorphism has the form $G \ni g \mapsto f^{-1}((h^{-1}(gh))f) \in G$, where f and h are fixed in G.

From conditions (3.85) and (3.86) we deduce that there is $n \in \{0, 1, 2, \cdots, \infty\}$ such that

$$card(\Upsilon(n)) = card(\Upsilon_\infty(n)) = card(\Upsilon), \tag{3.88}$$

where $\Upsilon(n) := \{b \in \Upsilon : ord(b) = n\}$.

From conditions (3.86)-(3.88) it follows that there are automorphisms of $gr_J(\Upsilon(n))$ and of $gr_{J_\infty}(\Upsilon_\infty(n))$ which are not inner and induced by bijective surjective mappings of the sets $\Upsilon(n)$ and $\Upsilon_\infty(n)$ since $card(gr_J(\Upsilon(n))) = card(gr_J(\Upsilon)) = card(J) =: \xi \geq \aleph_0$ and $card(Aut(gr_J(\Upsilon(n)))) = 2^\xi > \xi$. In virtue of the theorem about extensions of automorphisms [42] (or see [43, 44]) these automorphisms have extensions from the aforementioned subgroups on groups isomorphic with J and J_∞ respectively.

This implies that nontrivial outer automorphisms θ' of J and θ_∞' of J_∞ exist since $card(J \setminus gr_J(\Upsilon)) \leq card(J)$. They induce outer automorphisms θ of G and θ_∞ of H correspondingly since F_o is the commutative multiplicative normal subgroup in them. From their construction it follows that their restrictions on

F_o is the identity mapping $\theta|_{F_o} = id$ and $\theta_\infty|_{F_o} = id$. On the other hand, $G \cap T = F_o$, consequently, there exist T-linear extensions of automorphisms θ and θ_∞ from metagroups G and H on metagroup algebras $T[G]$ and $T[H]$ respectively.

The cardinality of the family of all outer T-linear automorphisms of A is not less, than 2^ξ since $2^\xi > \xi \geq \aleph_0$.

Corollary 3.3. Let conditions of Theorem 3.9 be satisfied.

Moreover, suppose that G is a topological Hausdorff metagroup and for J in the quotient topology $\Upsilon(n)$ fulfilling conditions (3.85)-(3.88)

$$\text{is everywhere dense in } J. \tag{3.89}$$

Then the metagroup G has discontinuous outer automorphisms.

If $A = T[G]$ is a topological Hausdorff metagroup algebra such that conditions in (3.89)

$$\text{are fulfilled for } G \text{ in the hereditary from } A \text{ topology,} \tag{3.90}$$

then A has discontinuous outer T-linear automorphisms.

Proof. (3.89). Take any two nonintersecting open subsets $U_1 = U_1^{-1}$ and U_2 in J with $U_2 = gU_1$ for some $g \in J$ and subsets $V_{i,j} \subset U_i$ such that $V_{i,1} \cap V_{i,2} = \emptyset$ and $V_{i,1} \cup V_{i,2} \subseteq \Upsilon(n) \cap U_i$, $\text{card}(V_{1,1}) = \text{card}(V_{i,j})$ and $V_{i,j}$ is everywhere dense in U_i for each i and j in $\{1, 2\}$. Using conditions (3.85)-(3.88) choose an automorphism θ' of J having the property $\theta'(V_{i,1}) = V_{i,1}$ for $i = 1$ and $i = 2$, $\theta'(V_{1,2}) = V_{2,2}$ and $\theta'(V_{2,2}) = V_{1,2}$. Thus θ' is discontinuous. It induces a discontiuous automorphism θ of G. (3.90) follows from (3.89).

Remark 3.3. For Banach algebras A and B over a normed field T such that T is complete relative to its norm uniformity by $A \hat{\otimes}_T B$ is denoted a completion of the tensor product $A \otimes_T B$ over the field T such that $A \hat{\otimes}_T B$ is also a Banach algebra into which A and B have natural emebeddings, for example, a completion relative to the projective tensor product topology (see [45–47]). For infinite fields with non-archimedean multiplicative norms and of zero characteristic the exponential and the logarithmic functions are described in [48] and on the Cayley-Dickson algebra $A_\infty(1)$ and on its l_2 completion $A_{\infty,2}(1)$ over $\mathbf{R}$ in [24, 49] (with $f_n = 1$ for each n).

Theorem 3.10. Let conditions of Theorem 3.9 be satisfied and let $\mathcal{T}$ be an infinite field complete relative to its multiplicative nontrivial norm with values in $[0, \infty)$, $\mathrm{char}(\mathcal{T}) = 0$. Let a metagroup algebra $A = \mathcal{T}[G]$ be normed and $\tilde{A}$ be its norm completion and $\tilde{A}$ be power-associative and the continuous logarithmic function from $\tilde{A} \setminus \{0\}$ into $\tilde{A}$ exist. Let also for each $g \in J$ there exists $u_g \in gF_\circ \subset G$ such that $|u_g| = 1$ and $\inf\{|u_g - u_h| : g \neq h \in J\} > 0$. Then a family $\mathrm{Out}^c_{\mathcal{T}}(\tilde{A}, \tilde{A}^e)$ of all continuous $\mathcal{T}$-linear derivations has the cardinality $\mathrm{card}(\mathrm{Out}^c_{\mathcal{T}}(\tilde{A}, \tilde{A}^e)) \geq 2^\xi \cdot \mathrm{card}(\mathcal{T})$, where $\tilde{A}^e = \tilde{A} \hat{\otimes}_{\mathcal{T}} \tilde{A}^{op}$.

Proof. Take elements h_g of G which are representatives of classes $gF_\circ$ and take $p_g \in F_\circ$ such that $|u_g| = 1$, where $u_g = p_g h_g$. They exist according to suppositions of this theorem. Then the $\mathcal{T}$-linear span of their family $\{u_g : g \in J\}$ is dense in $\tilde{A}$. Therefore, up to isomorphisms of metagroup algebras it is possible to take $\Upsilon(n) \subset \{u_g : g \in J\}$ and $\Upsilon_\infty(n) \subset \{u_g : g \in J_\infty\}$. Consider an outer automorphism θ of A constructed in the proof of Theorem 3.9 induced by bijective surjective mappings of $\Upsilon(n)$ and $\Upsilon_\infty(n)$. Since $|u_g| = 1$ for each $g \in J$ and $\inf\{|u_g - u_h| : g \neq h \in J\} > 0$, then $|\theta| < \infty$ and $|\theta^{-1}| < \infty$ and consequently, θ and θ^{-1} have continuous bijective extensions on $\tilde{A}$.

The algebra $\tilde{A}$ is power-associative, that is $a^n a^m = a^{n+m}$ for every $a \in A$ and natural numbers n and m, where $a^1 = a$, $a^n = a^{n-1} a$ for each $n \geq 2$, also $a^0 = 1$ for $a \neq 0$. Therefore, $\ln \theta : \tilde{A} \setminus \{0\} \to \tilde{A}$ is the continuous function and $\theta = e^{\ln \theta}$ since $a = e^{\ln a}$ and $\ln a \in \tilde{A}$ for each $a \in \tilde{A} \setminus \{0\}$.

A field $\mathcal{T}$ is infinite and with a multiplicative nontrivial norm with values in $[0, \infty)$, consequently, 0 is the limit point of $\Gamma_{\mathcal{T}} := \{|b| : b \in \mathcal{T} \setminus \{0\}\}$ in $\mathbb{R}$. Let $\eta : G \to H_\infty$ be an isomorphism of metagroups. It induces a $\mathcal{T}$-linear isomorphism of Banach algebras $\eta : \tilde{A} \to \tilde{A}_\infty$, where $\tilde{A}_\infty = \mathcal{T}[\tilde{H}_\infty]$.

For the local one-parameter group $\{e^{b \ln \theta} : b \in \mathcal{T}, |b| < \delta\}$ with a fixed $0 < \delta < \infty$ we get $\frac{de^{b \ln \theta}}{db} = e^{b \ln \theta} \ln \theta$ since the field $\mathcal{T}$ is embedded into $\tilde{A}$ as $\mathcal{T}1$. That is, $\ln \theta$ is the generator of the local one-parameter group $\theta^b = e^{b \ln \theta}$ acting from $\tilde{A}$ into $\tilde{A}$ such that $\theta^{b+c} = \theta^b \theta^c$ for each $|b| < \delta$, $|c| < \delta$, and $|b+c| < \delta$ in $\mathcal{T}$ with $\theta^0 = id$ since the field $\mathcal{T}$ is contained in the center of the metagroup algebra $\tilde{A}$. Considering the local one-parameter groups $\{e^{b \ln \theta} : b \in \mathcal{T}, |b| < \delta\}$ and $\{e^{b \ln x} : b \in \mathcal{T}, |b| < \delta\}$ for arbitrary $x \in \tilde{A} \setminus \{0\}$ one gets $dt = 0$ for each $t \in \mathcal{T}$ since θ is $\mathcal{T}$-linear and $\theta(0) = 0$, where $dx = \lim_{b \to 0} \frac{d\theta^b(x)}{db}$.

The algebra $\tilde{A}$ supplied with multiplication operations of anticommutation and commutation induced by the initial one such that $\{x, y\} := xy + yx$ and $[x, y] := xy - yx$ for each x and y in $\tilde{A}$ we denote by A^+ and A^- correspondingly. Therefore $\theta^b : \tilde{A}^+ \to \tilde{A}^+$ and $\theta^b : \tilde{A}^- \to \tilde{A}^-$ is such that $\theta^b\{x,y\} = \{\theta^b(x), \theta^b(y)\}$ and $\theta^b[x,y] = [\theta^b(x), \theta^b(y)]$ for every x and y in $\tilde{A}$ and $|b| < \delta$ in $\mathcal{T}$ since $\{x,y\} = \{y,x\}$ and $[x,y] = -[y,x]$ and $(-1)^b = e^{b\ln(-1)} \in \mathcal{T}$ and θ is the $\mathcal{T}$-linear automorphism of $\tilde{A}$. Hence

$$\frac{d\theta^b\{x,y\}}{db}\Big|_{b=0} = \{dx, y\} + \{x, dy\}, \text{ and}$$

$$\frac{d\theta^b[x,y]}{db}\Big|_{b=0} = [dx, y] + [x, dy],$$

for each x and y in $\tilde{A}$. On the other hand, $\lim_{b\to 0} \frac{d[e^{bu}e^{bz} - e^{b(u+z)}]}{db} = 0$ for all u and z in $\tilde{A}$ and consequently, $d(x+y) = dx + dy$ for each x and y in $\tilde{A}$. From $2xy = \{x,y\} + [x,y]$, it follows that $d(xy) = (dx)y + x(dy)$ for each x and y in $\tilde{A}$ since $\text{char}(\mathcal{T}) \neq 2$. Thus $\delta = (s\eta + r)d$ is a continuous derivation of the metagroup algebra $\tilde{A}$ which is noninner for each s and r in $\mathcal{T}$ with $|s| + |r| > 0$ since $\theta = e^d$ is the noninner automorphism.

The family of all automorphisms of the aforementioned type θ which provide such continuous derivations δ is not less than 2^ξ, consequently, $\text{card}(\text{Out}^c_{\mathcal{T}}(\tilde{A}, \tilde{A}^e)) \geq 2^\xi \cdot \text{card}(\mathcal{T})$ since $2^\xi > \xi \geq \aleph_0$ and $\xi \times \aleph_0 = \xi$ and $\text{card}(\mathcal{T}) \geq \aleph_0$.

Proposition 3.3. Let A be a nonassociative Banach algebra over an infinite field $\mathcal{T}$ complete relative to its multiplicative norm of $\text{char}(\mathcal{T}) = 0$ and let θ be an outer automorphism of A such that $|1 - \theta| < 1$. Then there exists an outer continuous derivation d of A such that $e^d = \theta$.

Proof. This follows from the proof of Theorem 3.10. Putting

$$d = \ln \theta = \ln(I - (I - \theta)) = -\sum_{n=1}^{\infty} \frac{(I - \theta)^n}{n},$$

since the series on the right is norm convergent.

Proposition 3.4. Suppose that A is the Cayley-Dickson algebra either $A_\infty(f_n)$ with $A_0 = \mathcal{T}$ over either (i) $\mathcal{T} = \mathbf{R}$ or (ii) an infinite (commutative) field $\mathcal{T}$ complete relative to its multiplicative norm of $\text{char}(\mathcal{T}) = 0$ and such that a natural number $l \in \mathbf{N}$ exists for which $\text{card}(n : f_n = f_l, n \in \mathbf{N}) = \aleph_0$ or (iii) in addition to the condition either (i) or (ii) let A be its one of the norm completions. Then the cardinality of a family of all $\mathcal{T}$-linear (continuous) outer derivations of A is not less than $2^{\aleph_0} \cdot \text{card}(\mathcal{T})$.

Proof. In case (i), the Cayley-Dickson algebra $A_\infty(f_n)$ over $\mathbf{R}$ is isomorphic with the Cayley-Dickson algebra $A_\infty(g_n)$ such that $g_n \in \{-1, 1\}$ for each $n \in \mathbf{N}$. Consequently, up to an isomorphism, card$(n : f_n = f_l, n \in \mathbf{N}) = \aleph_0$ with either $f_l = 1$ or $f_l = -1$.

For the Cayley-Dickson subalgebra $A(g)$ of A with $g_n = f_l$ for each $n \in \mathbf{N}$ denotes basic generators of $A(g)$ by $i_0, i_1, i_2, \cdots$ with $i_0 = 1$ and we take $G = \{i_0, i_1, i_2, \cdots\} \times \mathcal{T}_\circ$ with $\mathcal{T}_\circ$ being the multiplicative group of $\mathcal{T} \setminus \{0\}$. If A is normed, then we consider $A(g)$ as the closure of $A_\infty(g)$ in A. Then we put H to be generated by all finite multiplications of the doubling generators $i_{2^{2n}}$ and $\mathcal{T}_\circ$. Therefore, H is the normal proper submetagroup of G such that card$(G/H) = \aleph_0$.

Next, we consider a $\mathcal{T}$-linear automorphism θ of $A_\infty(g)$ first defined on doubling generators and then on their finite products since $A_\infty(g)$ and its norm completion has a basis of generators over the field $\mathcal{T}$. Let $P_1 \cup P_2 = \{1, 2, 3, \cdots\} = \mathbf{N}$ with $P_1 \cap P_2 = \emptyset$ and let $w : P_1 \to P_2$ be a bijective surjective map. Then we put

$$\theta(i_{2^r}) = \alpha_r i_{2^r} + \beta_r i_{2^m}, \text{ and}$$
$$\theta(i_{2^m}) = -\beta_r i_{2^r} + \alpha_r i_{2^m},$$

with $m = w(r)$, $\alpha_r^2 + \beta_r^2 = 1$ and $0 < |\beta_r| < 1/4$ for each $r \in P_1$, with $\theta(i_1) = i_1$ and $\theta(1) = 1$. Hence

$$\theta(i_{2^r})\theta(i_{2^m}) = i_{2^r} i_{2^m} = \theta(i_{2^r} i_{2^m}),$$
$$(\theta(i_{2^r}))^2 = (i_{2^r})^2, \text{ and}$$
$$(\theta(i_{2^m}))^2 = (i_{2^m})^2,$$

for each $r \in P_1$ with $m = w(r)$ since $i_k i_j = -i_j i_k$ for each $1 \le k$ and $1 \le j$ so that $j \ne k$, also $g_n = f_l = g_1$ for each $n \in \mathbf{N}$. Moreover, the map θ preserves the quadratic form $\bar{z}z$ for each $z \in A(g)$ and consequently, $\theta(t) = t$ for all t in the field $\mathcal{T}$.

Suppose that δ is an inner $\mathcal{T}$-linear continuous derivation of $A(g)$, that is realized as $\delta y = [m, y]$ for each $y \in A(g)$, where m is fixed in the enveloping algebra $A^e(g) = A(g) \hat{\otimes}_\mathcal{T} A^{op}(g)$. For each $y \in A(g)$ the limit $\delta y = \lim_{n\to\infty} \delta_n \pi_n y$ exists, where $\pi_n : A(g) \to A_n(g_{(n)})$ is the $\mathcal{T}$-linear projection from $A(g)$ as the Banach space on its subspace so that it induces the projection $\pi_n : A^e(g) \to A_n^e(g_{(n)})$ since $A(g)$ has the basis $\{i_0, i_1, i_2, \cdots\}$, also $[\pi_n m, \pi_n y] = \delta_n(\pi_n y)$ such that

$$\delta|_{A_n(g_{(n)})} = \delta_n, \quad \delta_n : A_n(g_{(n)}) \to A_n(g_{(n)}).$$

Then $\lim_{n\to\infty} \delta_n = \delta$ in operator norm since $\delta_n = \mathrm{ad}(\pi_n m)$. Then

$$\delta_n = [L_x, L_z] + [L_x, R_z] + [R_x, R_z],$$

for each $n \geq 3$, where $x \in \rho_n(A_3(g_{(3)}))$ and $z \in (\rho_n(A_3(g_{(3)})))^{op}$. Also

$$\rho_n : A_3(g_{(3)}) \hookrightarrow A_n(g_{(n)}),$$

is a $\mathcal{T}$-linear embedding (see Example 3.4 and references therein), $A_n(g_{(n)}) \subset A(g)$. Considering operators δ_3, $\delta_4 - \delta_3$, and $\delta_{n+1} - \delta_n$ for each $n \geq 4$, we deduce that an embedding $\psi : A_4(g_{(4)}) \hookrightarrow A(g)$ exists such that $\delta = \mathrm{ad}(m)$ for some $m \in (\psi(A_4(g_{(4)})))^e$ since $\psi = \lim_n \psi_n$, where ψ_n is induced by $\rho_n \oplus i_8 \rho_{n+1}$ using the isomorphism $A_3(g_{(3)}) \oplus i_8 A_3(g_{(3)})$ with $A_4(g_{(4)})$.

If we choose $0 < \sum_{r=1}^{\infty} |\beta_r| < 1/5$ in the real case, $\max_{1 \leq r < \infty} |\beta_r| < 1/4$ in the non-archimedean case, then $|1 - \theta| < 1$, consequently, θ has an extension to a continuous $\mathcal{T}$-linear automorphism of $A(g)$. In view of Proposition 3.3, the automorphism θ induces a continuous $\mathcal{T}$-linear derivation d of $A(g)$.

Each such automorphism is outer if $\mathrm{card}\{r : \beta_r \neq 0\} = \aleph_0$ since $\theta = e^d$ and d is outer because it can not be realized with the help of a finite number of multiplications and additions in $A(g)$. The family of such outer automorphisms of $A(g)$ has the cardinality $2^{\aleph_0} \cdot \mathrm{card}(\mathcal{T}) \geq 2^{\aleph_0}$.

These derivations can be extended onto the entire Cayley-Dickson algebra A since $A(g)$ has embedding into A. For example, this can be implemented by putting $\theta(1) = 1$ for the automorphism extension and hence $\delta(1) = 0$ for the corresponding derivation for each doubling generator 1 of A which is not in the list $\{i_{2^n} : n \in \mathbb{N}\}$ of the subalgebra $A(g)$. Then we extend θ by the multiplication rule and $\mathcal{T}$-linearity on A since $\theta|_{\mathcal{T}} = id$. Moreover, $|1 - \theta| < 1$ according to the construction of θ. This automorphism θ of A is outer and the corresponding to it derivation d is outer as follows from the proof above.

Thus by virtue of Theorems 3.9, 3.10 and Proposition 3.3, the Cayley-Dickson algebra $A(g)$ or its norm completion has not less than $2^{\aleph_0} \cdot \mathrm{card}(\mathcal{T})$ outer $\mathcal{T}$-linear (continuous) derivations.

Remark 3.4. For nonassociative algebras A over the real field $\mathbf{R}$ one defines von Neumann algebras and C^*-algebras (see [34]) quite analogously to that of associative algebras over $\mathbf{C}$ but with the difference that $t^* = t$ for each scalar t in $\mathbf{R}$, where $\mathbf{R}$ is contained in the center of A. Evidently, each C^*-algebra over $\mathbf{C}$ has also a C^*-algebra structure over $\mathbf{R}$.

For comparison it is worth to mention that for associative von Neumann algebras and simple C^*-algebras (operator algebras acting on Hilbert spaces) over the complex field all $\mathbf{C}$-linear derivations are continuous and inner (see [34, 35]).

Proposition 3.5. There exist nonassociative von Neumann algebras and simple C^*-algebras A, also semisimple C^*-algebras, which have families $\mathrm{Out}_{\mathbf{R}}^c(A, A^e)$ of outer continuous $\mathbf{R}$-linear derivations of the cardinality $\mathrm{card}(\mathrm{Out}_{\mathbf{R}}^c(A, A^e)) \geq 2^{\aleph_0}$.

Proof. The Cayley-Dickson algebra $A = A_{\infty,2}(f)$ over $\mathbf{R}$ with $f_n = 1$ for each n (see Theorems 2.1 and 2.5) is an example of the nonassociative analog of a C^*-algebra in which the norm is $|x| = \sqrt{\bar{x}x}$ and the scalar product is $(x, y) = \mathrm{Re}(\bar{x}y)$ for each x and y in A, where $x^* = \bar{x}$. From the multiplication rule in A it follows that it does not contain any proper nonzero (closed) ideal, hence it is simple. Elements x of the Cayley-Dickson algebra A also act as operators on the Hilbert two-sided module $M = A$. In view of Proposition 3.4, we deduce that $\mathrm{card}(\mathrm{Out}_{\mathbf{R}}^c(A, A^e)) \geq 2^{\aleph_0}$, where $A^e = A \hat{\otimes}_{\mathbf{R}} A^{op}$.

If B is either a von Neumann algebra or a semisimple C^*-algebra, then their direct sum $A \oplus B = E$ is such and $\mathrm{card}(\mathrm{Out}_{\mathcal{T}}^c(E, E^e)) \geq 2^{\aleph_0}$.

On the other hand, the completion of the tensor product $S = A \hat{\otimes}_{\mathbf{R}} B$ relative to the norm $|z| = (\sum_{k=0}^{\infty} |z_k|^2)^{1/2}$ is a von Neumann algebra or a semisimple C^*-algebra respectively, where $z \in S$, $z = \sum_{k=0}^{\infty} i_k z_k$, $z_k \in B$ for each k. This S is the left A-algebra. Similarly is constructed for the right A-algebra. A two-sided A-algebra is obtained from S by imposing the condition $zu = uz$ for each $u \in A$ and $z \in B$, when $A \cap B = \mathbf{R}$. Evidently, $\mathrm{Der}_{\mathbf{R}}^c(A, A^e)$ provides an infinite dimensional Lie algebra $\mathcal{L}(\mathrm{Der}_{\mathbf{R}}^c(A, A^e))$ with the Lie multiplication $[\delta_1, \delta_2] = \delta_1 \delta_2 - \delta_2 \delta_1$ of derivations δ_1 and δ_2 considered as operators on A, where $\mathrm{Der}_{\mathbf{R}}^c(A, A^e)$ is the family of all continuous $\mathbf{R}$-linear derivations of A.

If B is finite dimensional or $\mathrm{Out}_{\mathbf{R}}^c(A, A^e) \setminus B$ is infinite dimensional over $\mathbf{R}$, then $\mathrm{card}(\mathrm{Out}(S, S^e)) \geq 2^{\aleph_0}$. The latter is accomplished particularly for commutative B since from the proof of Proposition 3.4, we infer that in $\mathrm{Out}_{\mathbf{R}}^c(A, A^e)$ there is a family Ω of pairwise noncommuting elements such that $\mathrm{card}\{\delta \in \Omega : \delta \notin \mathcal{L}(\Omega \setminus \delta)\} \geq 2^{\aleph_0}$, where $\mathcal{L}(\Psi)$ denotes a minimal Lie algebra generated by Ψ.

Proposition 3.6. Let conditions of Theorem 3.10 be fulfilled and B be a Banach algebra over $\mathcal{T}$ such that $\mathrm{Out}^c_{\mathcal{T}}(\tilde{A}, \tilde{A}^e) \setminus B$ is infinite dimensional over $\mathcal{T}$. Then $\mathrm{card}(\mathrm{Out}^c_{\mathcal{T}}(E, E^e)) \geq 2^\xi \cdot \mathrm{card}(\mathcal{T})$, where $E = A \hat{\otimes}_{\mathcal{T}} B$.

Corollary 3.4. If conditions of Theorem 3.10 are satisfied and a unital Banach algebra B over $\mathcal{T}$ is commutative, $E = A \hat{\otimes}_{\mathcal{T}} B$, then $\mathrm{card}(\mathrm{Out}^c_{\mathcal{T}}(E, E^e)) \geq 2^\xi \cdot \mathrm{card}(\mathcal{T})$ and $\mathrm{card}(\mathrm{Out}^c_B(E, E^e)) \geq 2^\xi \cdot \mathrm{card}(\mathcal{T})$.

The proofs of Proposition 3.6 and Corollary 3.4 are similar to that of Proposition 3.5.

4. APPENDIX: GENERALIZED HYPERCOMPLEX NUMBERS

The main subject of this chapter are structure of hypercomplex Hilbert modules, derivations of operator algebras on hypercomplex Hilbert spaces and related cohomologies. Nonetheless, in this section it is shortly demonstrated that there are abundant families of metagroups besides those which appear in areas described in the introduction. Then metagroup algebras provide algebras of hypercomplex numbers.

Theorem 4.1. Let G_j be a family of metagroups (see Definition 3.1), where $j \in J$, J is a set. Then their direct product $G = \prod_{j \in J} G_j$ is a metagroup and

$$\mathcal{C}(G) = \prod_{j \in J} \mathcal{C}(G_j). \tag{A.1}$$

Proof. Each element $a \in G$ is written as $a = \{a_j : \forall j \in J, a_j \in G_j\}$. Therefore, a product

$$ab = \{c : \forall j \in J, c_j = a_j b_j, a_j \in G_j, b_j \in G_j\},$$

is a single-valued binary operation on G. Then we get that

$$a \setminus b = \{d : \forall j \in J, d_j = a_j \setminus b_j, a_j \in G_j, b_j \in G_j\}, \text{ and}$$
$$a/b = \{d : \forall j \in J, d_j = a_j/b_j, a_j \in G_j, b_j \in G_j\}.$$

Moreover, $e_G = \{\forall j \in J, e_{G_j}\}$ is a neutral element in G, where e_{G_j} denotes a neutral element in G_j for each $j \in J$. Thus conditions (3.1)-(3.3) are satisfied.

From Conditions (3.4)-(3.7) for each G_j in Definition 3.1, we infer that

$$\mathrm{Com}(G) := \{a \in G : \forall b \in G, ab = ba\} =$$
$$\left\{ a \in G : a = \{a_j : \forall j \in J, a_j \in G_j\}; \right.$$
$$\forall b \in G, b = \{b_j : \forall j \in J, b_j \in G_j\};$$
$$\left. \forall j \in J, a_j b_j = b_j a_j \right\}$$
$$= \prod_{j \in J} \mathrm{Com}(G_j), \qquad (A.2)$$

$N_l(G) := \{a \in G : \forall b \in G, \forall c \in G, (ab)c = a(bc)\} = \{a \in G : a = \{a_j : \forall j \in J, a_j \in G_j\}; \forall b \in G, b = \{b_j : \forall j \in J, b_j \in G_j\}; \forall c \in G, c = \{c_j : \forall j \in J, c_j \in G_j\}; \forall j \in J, (a_j b_j) c_j = a_j (b_j c_j)\} = \prod_{j \in J} N_l(G_j),$ (A.3)

and similarly

$$N_m(G) = \prod_{j \in J} N_m(G_j), \qquad (A.4)$$

and

$$N_r(G) = \prod_{j \in J} N_r(G_j). \qquad (A.5)$$

This and (3.8) imply that

$$N(G) = \prod_{j \in J} N(G_j). \qquad (A.6)$$

Thus

$$\mathcal{C}(G) := \mathrm{Com}(G) \cap N(G) = \prod_{j \in J} \mathcal{C}(G_j). \qquad (A.7)$$

Let a, b and c be in G, then

$$(ab)c = \{(a_j b_j) c_j : \forall j \in J, a_j \in G_j, b_j \in G_j, c_j \in G_j\}$$
$$= \{\mathrm{t}_{3,G_j}(a_j, b_j, c_j) a_j (b_j c_j) : \forall j \in J, a_j \in G_j, b_j \in G_j, c_j \in G_j\}$$
$$= \mathrm{t}_{3,G}(a, b, c) a(bc),$$

where

$$t_{3,G}(a,b,c) = \{t_{3,G_j}(a_j, b_j, c_j) : \forall j \in J,\, a_j \in G_j, b_j \in G_j, c_j \in G_j\}. \tag{A.8}$$

Therefore, formulas (A.7) and (A.8) imply that condition (3.9) also is satisfied. Thus G is a metagroup.

Remark 4.1. Let A and B be two metagroups and let $\mathcal{C}$ be a commutative group such that

$$\mathcal{C}_m(A) \hookrightarrow \mathcal{C},\ \mathcal{C}_m(B) \hookrightarrow \mathcal{C},\ \mathcal{C} \hookrightarrow \mathcal{C}(A) \text{ and } \mathcal{C} \hookrightarrow \mathcal{C}(B), \tag{A.9}$$

where $\mathcal{C}_m(A)$ denotes a minimal subgroup in $\mathcal{C}(A)$ generated by

$$\{t_A(a,b,c) : a \in A, b \in A, c \in A\}.$$

Using direct products it is always possible to extend either A or B to get such a case. In particular, either A or B may be a group. Let on the Cartesian product $A \times B$ an equivalence relation Ξ be such that

$$(\gamma v, b)\Xi(v, \gamma b) \text{ and } (\gamma v, b)\Xi \gamma(v, b) \text{ and } (\gamma v, b)\Xi(v, b)\gamma, \tag{A.10}$$

for every v in A, b in B and γ in $\mathcal{C}$.

Let $\phi : A \to \mathcal{A}(B)$ be a single-valued mapping, where $\mathcal{A}(B)$ denotes a family of all bijective surjective single-valued mappings of B onto B subjected to the conditions given below. If $a \in A$ and $b \in B$, then it will be written shortly b^a instead of $\phi(a)b$, where $\phi(a) : B \to B$. Let also

$$\eta_\phi : A \times A \times B \to \mathcal{C},\ \kappa_\phi : A \times B \times B \to \mathcal{C},\ \text{and}$$
$$\xi_\phi : ((A \times B)/\Xi) \times ((A \times B)/\Xi) \to \mathcal{C}$$

be single-valued mappings written shortly as η, κ and ξ correspondingly such that

$$(b^u)^v = b^{vu}\eta(v, u, b),\ e^u = e,\ b^e = b; \tag{A.11}$$
$$\eta(v, u, \gamma b) = \eta(v, u, b); \tag{A.12}$$
$$(cb)^u = c^u b^u \kappa(u, c, b); \tag{A.13}$$
$$\kappa(u, \gamma c, b) = \kappa(u, c, \gamma b) = \kappa(u, c, b) \text{ and } \kappa(u, \gamma, b) = \kappa(u, b, \gamma) = e; \tag{A.14}$$
$$\xi((\gamma u, c), (v, b)) = \xi((u, c), (\gamma v, b)) = \xi((u, c), (v, b)); \tag{A.15}$$
$$\xi((\gamma, e), (v, b)) = e \text{ and } \xi((u, c), (\gamma, e)) = e, \tag{A.16}$$

for every u and v in A, b, c in B, γ in C, where e denotes the neutral element in C and in A and B.

We put

$$(a_1, b_1)(a_2, b_2) = (a_1 a_2, \xi((a_1, b_1), (a_2, b_2)) b_1 b_2^{a_1}), \quad (A.17)$$

for each a_1, a_2 in A, b_1 and b_2 in B.

The Cartesian product $A \times B$ supplied with such a binary operation (A.17) will be denoted by $A \otimes^{\phi, \eta, \kappa, \xi} B$.

Theorem 4.2. Let the conditions of Remark 4.1 be fulfilled. Then the Cartesian product $A \times B$ supplied with a binary operation (A.17) is a metagroup.

Proof. From the conditions of Remark 4.1 it follows that the binary operation (A.17) is single-valued.

Let

$$I_1 = ((a_1, b_1)(a_2, b_2))(a_3, b_3) \text{ and } I_2 = (a_1, b_1)((a_2, b_2)(a_3, b_3)),$$

where a_1, a_2, a_3 belong to A, where b_1, b_2, b_3 belong to B. Then we infer that

$$I_1 = \Big((a_1 a_2) a_3, \xi((a_1, b_1), (a_2, b_2)) \xi((a_1 a_2, b_1 b_2^{a_1}), (a_3, b_3))(b_1 b_2^{a_1}) b_3^{a_1 a_2}\Big), \text{ and}$$

$$I_2 = \Big(a_1 (a_2 a_3), \xi((a_1, b_1), (a_2 a_3, b_2 b_3^{a_2})) [\xi((a_2, b_2), (a_3, b_3))]^{a_1}$$

$$b_1 (b_2^{a_1} b_3^{a_1 a_2}) \kappa(a_1, b_2, b_3^{a_2}) \eta(a_1, a_2, b_3)\Big).$$

Therefore

$$I_1 = \mathsf{t}_3((a_1, b_1), (a_2, b_2), (a_3, b_3)) I_2, \quad (A.18)$$

with

$$\mathsf{t}_3((a_1, b_1), (a_2, b_2), (a_3, b_3)) = \mathsf{t}_{3,A}(a_1, a_2, a_3) \mathsf{t}_{3,B}(b_1, b_2^{a_1}, b_3^{a_1 a_2})$$
$$\xi((a_1, b_1), (a_2 a_3, b_2 b_3^{a_2})) [\xi((a_2, b_2), (a_3, b_3))]^{a_1} \kappa(a_1, b_2, b_3^{a_2}) \eta(a_1, a_2, b_3)$$
$$/ [\xi((a_1, b_1), (a_2, b_2)) \xi((a_1 a_2, b_1 b_2^{a_1}), (a_3, b_3))]. \quad (A.19)$$

Apparently, $\mathsf{t}_{3, A \otimes^{\phi, \eta, \kappa, \xi} B}((a_1, b_1), (a_2, b_2), (a_3, b_3)) \in C$ for each $a_j \in A$, $b_j \in B$, $j \in \{1, 2, 3\}$, where for shortening of a notation $\mathsf{t}_{3, A \otimes^{\phi, \eta, \kappa, \xi} B}$ is denoted by t_3.

If $\gamma \in \mathcal{C}$, then

$$\gamma((a_1,b_1)(a_2,b_2)) = (\gamma a_1 a_2, \xi((a_1,b_1),(a_2,b_2))b_1 b_2^{a_1})$$
$$= (a_1 a_2, b_1 b_2^{a_1})\gamma\xi((a_1,b_1),(a_2,b_2))$$
$$= ((a_1,b_1)(a_2,b_2))\gamma.$$

Hence $\gamma \in \mathcal{C}(A \otimes^{\phi,\eta,\kappa,\xi} B)$, consequently, $\mathcal{C} \subseteq \mathcal{C}(A \otimes^{\phi,\eta,\kappa,\xi} B)$.

Next we consider the following equation

$$(a_1,b_1)(a,b) = (e,e), \qquad (\text{A.20})$$

where $a \in A, b \in B$.

From (3.2) and (A.17) we deduce that

$$a_1 = e/a. \qquad (\text{A.21})$$

Consequently, $\xi((e/a, b_1), (a,b))b_1 b^{(e/a)} = e$, and hence

$$b_1 = e/[\xi((e/a, b^{(e/a)}), (a,b))b^{(e/a)}]. \qquad (\text{A.22})$$

Thus $a_1 \in A$ and $b_1 \in B$ given by (A.21) and (A.22) provide a unique solution of (A.20).

Similarly from the following equation

$$(a,b)(a_2,b_2) = (e,e), \qquad (\text{A.23})$$

where $a \in A, b \in B$ we infer that

$$a_2 = a \setminus e. \qquad (\text{A.24})$$

Consequently, $\xi((a,b), (a \setminus e, b_2))bb_2^a = e$, and hence $b_2^a = [\xi((a,b), (a \setminus e, b_2))b] \setminus e$. On the other hand, $(b_2^a)^{e/a} = \eta(e/a, a, b_2)b_2$. Consequently,

$$b_2 = (b \setminus e)^{e/a}\{[(\xi((a,b), (a \setminus e, (b \setminus e)^{e/a}))]^{e/a}\eta(e/a, a, (b \setminus e)^{e/a})\}. \qquad (\text{A.25})$$

Thus formulas (A.24) and (A.25) provide a unique solution of (A.23).

Next we put $(a_1,b_1) = (e,e)/(a,b)$ and $(a_2,b_2) = (a,b) \setminus (e,e)$ and

$$(a,b) \setminus (c,d) = ((a,b) \setminus (e,e))(c,d)\mathsf{t}_3((e,e)/(a,b), (a,b), ((a,b) \setminus (e,e))(c,d))$$
$$/\mathsf{t}_3((e,e)/(a,b), (a,b), (a,b) \setminus (e,e)); \qquad (\text{A.26})$$
$$(c,d)/(a,b) = (c,d)((e,e)/(a,b))\mathsf{t}_3((e,e)/(a,b), (a,b), (a,b) \setminus (e,e))$$
$$/\mathsf{t}_3((c,d)(e/(a,b)), (a,b), (a,b) \setminus (e,e)), \qquad (\text{A.27})$$

and $e_G = (e, e)$, where $G = A \otimes^{\phi,\eta,\kappa,\xi} B$. Note that (3.3) follows from (A.15) and (A.17). Therefore properties (3.2)-(3.3) and (3.9) are fulfilled for $A \otimes^{\phi,\eta,\kappa,\xi} B$.

Definition 4.1. The metagroup $A \otimes^{\phi,\eta,\kappa,\xi} B$ provided by Theorem 4.2 we call a smashed product of metagroups A and B with smashing factors ϕ, η, κ and ξ.

Remark 4.2. From Theorems 4.1 and 4.2, it follows that taking nontrivial η, κ and ξ and starting even from groups with nontrivial $\mathcal{C}(G_j)$ or $\mathcal{C}(A)$ it is possible to construct new metagroups with nontrivial $\mathcal{C}(G)$ and ranges $\mathrm{t}_{3,G}(G,G,G)$ of $\mathrm{t}_{3,G}$ may be infinite.

With suitable smashing factors ϕ, η, κ and ξ and with nontrivial metagroups or groups A and B it is easy to get examples of metagroups in which $e/a \neq a \setminus e$ for an infinite family of elements a in $A \otimes^{\phi,\eta,\kappa,\xi} B$. Evidently smashed products (see Remark 4.1 and Theorem 4.2) are nonassociative generalizations of semidirect products.

Conclusion

The results of this chapter can be used for further studies of structures of operator algebras on hypercomplex Hilbert spaces, their derivations and cohomologies, structure of hypercomplex Hilbert modules. Besides these applications it is worthwhile to mention possible applications in mathematical coding theory and its technical applications [50–52], because frequently codes are based on binary systems and algebras. It can be also used for studies of spectra of operators, for example, partial differential operators having technical applications [34,53,54].

References

[1] Dickson L. E. (1975), *The collected mathematical papers*. Volumes 1–5. Chelsea Publishing Co., New York.

[2] Baez J. C. (2002), The octonions. *Bulletin of the AMS - American Mathematical Society*, 39, 145–205.

[3] Brown R. B. (1967), On generalized Cayley-Dickson algebras. *Pacific Journal of Mathematics*, 20, 415–422.

[4] Kantor I. L., Solodovnikov A. S. (1989), *Hypercomplex numbers.* Springer-Verlag, Berlin.

[5] Schafer R. D. (1966), *An introduction to nonassociative algebras.* Academic Press, New York.

[6] Culbert C. (2007), Cayley-Dickson algebras and loops. *Journal of Generalized Lie Theory and Applications*, 1, 1–17.

[7] Gürsey F., Tze C.-H. (1996), *On the role of division, Jordan and related algebras in particle physics.* World Scientific Publ. Co., Singapore.

[8] Krausshar R. S. (2004), *Generalized analytic automorphic forms in hypercomplex spaces.* Birkhäuser, Basel.

[9] Serôdio R. (2007), On octonionic polynomials. *Advances in Applied Clifford Algebras*, 17, 245–258.

[10] Bourbaki N. (2007, 2012), *Algebra, Chapters 1–3.* Springer-Verlag, Berlin.

[11] Emch G. (1963), Mèchanique quantique quaternionienne et Relativitè restreinte. *Helvetica Physica Acta*, 36, 739–788.

[12] Gentili G., Struppa D. C. (2010), Regular functions on the space of Cayley numbers. *Rocky Mountain Journal of Mathematics*, 40, 225–241.

[13] Ghiloni R., Perotti A. (2011), Slice regular functions on real alternative algebras. *Advances in Mathematics*, 226, 1662–1691.

[14] Gilbert J. E., Murray M. A. M. (1991), *Clifford algebras and Dirac operators in harmonic analysis.* Cambr. studies in advanced Mathem. 26, Cambridge Univ. Press, Cambridge.

[15] Girard P. R. (2007), *Quaternions, Clifford algebras and relativistic Physics.* Birkhäuscr, Basel.

[16] Goto M., Grosshans F. D. (1978), *Semisimple Lie algebras.* Marcel Dekker, Inc., New York.

[17] Gürlebeck K., Sprössig W. (1997), *Quaternionic and Clifford calculus for physicists and engineers.* John Wiley and Sons, Chichester.

[18] Gürlebeck K., Spróssig W. (1990), *Quaternionic analysis and elliptic boundary value problem.* Birkhäuser, Basel.

[19] Jacobson N. (1939), Cayley numbers and normal simple Lie algebras of type G. *Duke Mathematical Journal*, 5, 775–783.

[20] Lawson H. B., Michelson M. -L. (1989), *Spin geometry.* Princ. Univ. Press, Princeton.

[21] Ludkovsky S. V. (2009/2010), Wrap groups of connected fiber bundles: their structure and cohomologies. *International Journal of Mathematics, Game Theory and Algebra*, 19, 53–128; parallel publication in: (2009). *Lie Groups: New Research,* Ed. A.B. Canterra, Nova Science Publishers, Inc., New York.

[22] Ludkovsky S. V. (2013), Meta-invariant operators over Cayley-Dickson algebras and spectra. *Advances in Pure Mathematics*, 3, 41–69.

[23] Ludkovsky S. V. (2008), Algebras of operators in Banach spaces over the quaternion skew field and the octonion algebra. *Journal of Mathematical Sciences*, New York, 144, 4301–4366.

[24] Ludkovsky S. V. (2007), Differentiable functions of Cayley-Dickson numbers and line integration. *Journal of Mathematical Sciences*, New York, 141, 1231–1298.

[25] Ludkovsky S. V., Spróssig, W. (2010), Ordered representations of normal and super-differential operators in quaternion and octonion Hilbert spaces. *Advances in Applied Clifford Algebras*, 20, 321–342.

[26] Ludkovsky S. V., Spróssig, W. (2011), Spectral theory of super-differential operators of quaternion and octonion variables. *Advances in Applied Clifford Algebras*, 21, 165–191.

[27] Ludkovsky S. V., Spróssig, W. (2012), Spectral representations of operators in Hilbert spaces over quaternions and octonions, *Complex Variables and Elliptic Equations*, 57, 1301–1324.

[28] Oystaeyen F. van. (2000), *Algebraic geometry for associative algebras,* Series *Lecture Notes in Pure and Applied Mathematics*, 232. Marcel Dekker Inc., New York.

[29] McCrimmon K. (1985), Derivations and Cayley derivations of generalized Cayley-Dickson algebras. *Pacific Journal of Mathematics*, 117, 163–182.

[30] Eakin P., Sathaye A. (1990), On automoprhisms and derivations of Cayley-Dickson algebras. *Journal of Algebra*, 129, 263–278.

[31] Brown K. S. (1994), *Cohomology of groups.* Springer-Verlag, New York.

[32] Cartan H., Eilenberg S. (1956), *Homological algebra.* Princeton Univ. Press, Princeton, New Jersey.

[33] Hochschild G. (1946), On the cohomology theory for associative algebras. *Annals of Mathematics*, 47, 568-579.

[34] Kadison R. V., Ringrose J. R. (1983), *Fundamentals of the theory of operator algebras.* Acad. Press, New York.

[35] Kadison R. V., Ringrose J. R. (1967), Derivations and automorphisms of operator algebras. *Communications in Mathematical Physics*, 4, 32–63.

[36] Rosenber A., Zelinsky D. (1956), Cohomology of infinite algebras. *Transactions of the AMS - American Mathematical Society*, 82, 85–98.

[37] Ludkovsky S. V. (2015), C^*-algebras of meta-invariant operators in modules over Cayley-Dickson algebras. *Southeast Asian Bulletin of Mathematics*, 39, 625–684.

[38] Edwards R. E. (1965), *Functional analysis.* Holt, Rinehart and Winston, New York.

[39] Engelking R. (1989), *General topology.* Heldermann, Berlin.

[40] Narici L., Beckenstein E. (1985), *Topological vector spaces.* Marcel-Dekker Inc., New York.

[41] Kunen K. (1980), *Set theory.* North-Holland Publishing Co., Amsterdam.

[42] Neumann B. H., Neumann H. (1952), Extending partial endomorphisms of groups. *Proceedings of the London Mathematical Society*, 2, 337–348.

[43] Plotkin B. I. (1966), *Groups of automorphisms of algebraic systems.* Nauka, Moscow.

[44] *Endomorphisms and endomorphism semigroups of groups,* in: Focus on group theory research. Ed.: L. M. Ying, Puusemp, P. (2006), Nova Science Publishers Inc., New York, 27–57.

[45] Grothendieck A. (1955), Produits tensoriels topologiques et espaces nucleaires. *Memoirs of the American Mathematical Society,* 16.

[46] Put M. van der, Tiel, J. van. (1967), Espaces nucléaires non-archimédiens. *Indagationes Mathematicae,* 29, 556-561.

[47] Rooij A. C. M. van. (1978), *Non-Archimedean functional analysis.* Marcel Dekker Inc., New York.

[48] Schikhof W. H. (1984), *Ultrametric calculus.* Cambridge, Cambridge Univ. Press.

[49] Ludkovsky S. V. (2010), *Analysis over Cayley-Dickson numbers and its applications.* LAP Lambert Academic Publishing, Saarbrücken.

[50] Blahut R. E. (2003), *Algebraic codes for data transmission.* Cambridge Univ. Press, Cambridge.

[51] Magomedov Sh. G. (2017), Assessment of the impact of confounding factors in the performance information security, *Russian Technological J.,* 5, 47–56.

[52] Shum K. P., Ren X., Wang Y. (2014), Semigroups on semilattice and the constructions of generalized cryptogroups, *Southeast Asian Bulletin of Mathematics,* 38, 719–730.

[53] Dunford N., Schwartz J. C. (1966), *Linear operators.* J. Wiley and Sons, Inc., New York.

[54] Zaikin B. A., Bogadarov A. Yu., Kotov A. F., Poponov, P. V. (2016), Evaluation of coordinates of air target in a two-position range measurement radar, *Russian Technological J.,* 4, 65–72.

In: Hilbert Spaces: Properties and Applications
Editor: Le Bin Ho

ISBN: 978-1-53616-633-0
© 2020 Nova Science Publishers, Inc.

Chapter 5

ON ANALYTIC SOLUTIONS OF THE DRIVEN, 2-PHOTON AND TWO-MODE QUANTUM RABI MODELS

Yao-Zhong Zhang*
School of Mathematics and Physics,
The University of Queensland Brisbane, QLD, Australia

Abstract

By applying the Bogoliubov transformations and through the introduction of the Bargmann-Hilbert spaces, we obtain analytic representations of solutions to the driven Rabi model without $\mathcal{Z}_2$ symmetry, and the 2-photon and two-mode quantum Rabi models. In each case, the transcendental function is analytically derived whose zeros give in the energy spectrum of the model. The zeros can be numerically found by standard root-search techniques.

Keywords: quasi-exactly solvable systems, Rabi model, solutions of wave equations

1. INTRODUCTION

Spin-boson systems describe interactions between spins and harmonic oscillators (boson modes) and have played a prominent role in modeling the ubiquitous

---
*Corresponding Author's E-mail: yzz@maths.uq.edu.au.

matter-light interactions in modern physics. One of the well-known spin-boson systems is the quantum Rabi model. This model describes the interaction of a two-level atom with a harmonic mode of quantized electromagnetic fields. Due to the simplicity of its Hamiltonian, the Rabi model has served as the basis for understanding matter-light interactions and has a variety of applications ranging from quantum optics [1] to solid state semiconductor systems [2] and molecular physics [3]. It has also played a significant role in the novel research field of cavities and circuit quantum electrodynamics [4,5].

The Hamiltonian of the Rabi model with $\mathcal{Z}_2$ symmetry is given by

$$H_R = \omega b^\dagger b + \Delta\, \sigma_z + g\, \sigma_x (b^\dagger + b), \tag{1.1}$$

where g is the spin-boson interaction strength, σ_z, σ_x are the Pauli matrices describing the two atomic levels separated by energy difference 2Δ, and $b^\dagger$ (b) are creation (annihilation) operators of a boson mode with frequency ω. In the Bargmann space with the monomials $\left\{\frac{w^n}{\sqrt{n!}}\right\}$ as basis vectors, the boson creation and annihilation operators can be realized as $b^\dagger = w$, $b = \frac{d}{dw}$ [6]. In terms of the two-component wavefunction $\psi(w) = (\psi_+(w), \psi_-(w))^T$, the time-independent Schrödinger equation of the model gives a system of two coupled linear differential equations with rational coefficients. The system depends on a spectral parameter E. The wavefunction components are elements of Bargmann-Hilbert (BH) space of entire functions of one variable $w \in \mathbf{C}$ [7].

The scalar product of any two elements $f(w), g(w)$ in the BH space is given by

$$\langle f|g\rangle = \int \overline{f(w)}\, g(w)\, d\mu(w), \tag{1.2}$$

where $\overline{f(w)}$ is the complex conjugate of $f(w)$ and $d\mu(w) = \frac{1}{\pi} e^{-|w|^2}\, dx\, dy$ is the measure. The energy E belongs to the spectrum of the problem if and only if for this value of E the system of coupled differential equations (given by the time-independent Schrödinger equation) has entire solution [7].

Using the BH space approach, Braak [8] recently presented a pair of transcendental functions defined as infinite power series expansions with coefficients satisfying three-term recurrence relations, and argued that the spectrum of the Rabi model is given by the zeros of the transcendental functions. This theoretical progress has renewed the interest in the Rabi and related models [9–19].

In this work, we apply the Bogoliubov transformations and the BH approach to obtain analytic solutions of the driven Rabi model with broken $\mathcal{Z}_2$ symmetry and the 2-photon and two-mode quantum Rabi models. The driven Rabi model is a generalization of the $\mathcal{Z}_2$ symmetrical Rabi model [8]. It recently has been used in [20] to examine quantum thermalization. The 2-photon and two-mode Rabi models are phenomenological models describing a two-level atom interacting with 2 photons and 2 harmonic modes, respectively. They can be experimentally realized in circuit quantum electromagnetic systems [5], and have established applications in many research fields, including Rubidium atoms [21] and quantum dots [22, 23].

Previous studies mainly concerned the Rabi model and its simple variations and presumed the standard measure (given above) for the corresponding BH spaces. However, as seen in Sections 3 and 4 below, for the 2-photon and two-mode Rabi models, the standard measure is no longer appropriate for defining the scalar products of their BH spaces. This is because the scalar product defined with the standard measure is not consistent with the differential representations (3.6) and (4.6). In this work, we introduce new BH spaces for the two models and find the suitable measures needed to define their scalar products. This leads to the completely new criteria for entire (wave)functions, as can be seen from (3.10) and (4.10) [cf. (2.4) related to the standard measure]. These criteria are essential in the derivation of the analytic solutions presented here for the 2-photon and two-mode Rabi models.

2. DRIVEN QUANTUM RABI MODEL

The Hamiltonian of the driven quantum Rabi model is given by

$$H_{dR} = \omega b^\dagger b + \Delta \sigma_z + g \sigma_x (b^\dagger + b) + \delta \sigma_x, \tag{2.1}$$

where δ is the drive amplitude. After a canonical Bogoliubov transformation $b = a + \frac{\lambda}{\omega}$, $b^\dagger = a^\dagger + \frac{\lambda}{\omega}$, where λ is a real parameter, the Hamiltonian (2.1) becomes

$$\tilde{H}_{dR} = \omega a^\dagger a + \Delta \sigma_z + (g \sigma_x + \lambda)(a^\dagger + a) + \delta \sigma_x + \frac{2g\lambda}{\omega} \sigma_x + \frac{\lambda^2}{\omega}. \tag{2.2}$$

Using the Bargmann realization $a^\dagger = z$, $a = \frac{d}{dz}$, working in a representation defined by σ_x diagonal and choosing $\lambda = -g$, we can turn the transformed driven Rabi Hamiltonian into the matrix differential operator

$$\tilde{H}_{dR} = \begin{pmatrix} \omega z \frac{d}{dz} + \delta - \frac{g^2}{\omega} & \Delta \\ \Delta & \omega \left(z - \frac{2g}{\omega}\right) \frac{d}{dz} - 2gz - \delta + \frac{3g^2}{\omega} \end{pmatrix}. \tag{2.3}$$

The BH space is the space of entire functions $f(z)$ with inner product (1.2) and orthonormal basis $\left\{\frac{z^n}{\sqrt{n!}}\right\}$. So if $f(z) = \sum_{n=0}^{\infty} c_n z^n$, then its norm $\|f\| = \sqrt{\langle f|f\rangle}$ is given by

$$\|f\|^2 = \sum_{n=0}^{\infty} |c_n|^2 n! \tag{2.4}$$

and $f(z)$ is entire analytic function iff this sum converges [6].

In terms of two-component wavefunction $\varphi(z) = (\varphi_+(z), \varphi_-(z))^T$, the time-independent Schrödinger equation, $\tilde{H}_{dR}\, \varphi(z) = E\, \varphi(z)$, yields a system of a couple of differential equations

$$\left(\omega z \frac{d}{dz} + \delta - \frac{g^2}{\omega} - E\right) \varphi_+ + \Delta \varphi_- = 0, \tag{2.5}$$

$$\left[\omega\left(z - \frac{2g}{\omega}\right)\frac{d}{dz} - 2gz + \frac{3g^2}{\omega} - \delta - E\right] \varphi_- + \Delta \varphi_+ = 0. \tag{2.6}$$

Solutions to these equations must be analytic in the whole complex plane if E belongs to the spectrum of $\tilde{H}_{dR}$. So we are looking for solutions of the form

$$\varphi_+(z) = \sum_{n=0}^{\infty} \mathcal{R}_n^+(E) z^n, \quad \varphi_-(z) = \sum_{n=0}^{\infty} \mathcal{R}_n^-(E) z^n, \tag{2.7}$$

which converge in the entire complex plane and are elements of the BH space with inner product (1.2).

Substituting (2.7) into (2.5), we obtain

$$\mathcal{R}_n^+ = \frac{\Delta}{E - n\omega - \delta + \frac{g^2}{\omega}} \mathcal{R}_n^-. \tag{2.8}$$

Thus $\mathcal{R}_n^+$ is not analytic in E but has simple poles at

$$E = n\omega + \delta - \frac{g^2}{\omega}, \quad n = 0, 1, \cdots. \tag{2.9}$$

This gives one set of exact energies of the driven Rabi model.

Choosing $\lambda = -g$ in (2.2), we get

$$\tilde{H}_{dR} = \begin{pmatrix} \omega\left(z + \frac{2g}{\omega}\right)\frac{d}{dz} + 2gz + \delta + \frac{3g^2}{\omega} & \Delta \\ \Delta & \omega z \frac{d}{dz} - \delta - \frac{g^2}{\omega} \end{pmatrix}. \tag{2.10}$$

Let $\phi(z) = (\phi_+(z), \phi_-(z))^T$ be a two-component wavefunction. Then the time-independent Schrödinger equation in this case yields

$$\left[\omega\left(z + \frac{2g}{\omega}\right)\frac{d}{dz} + 2gz + \frac{3g^2}{\omega} + \delta - E\right]\phi_+ + \Delta\,\phi_- = 0, \quad (2.11)$$

$$\left(\omega z\frac{d}{dz} - \delta - \frac{g^2}{\omega} - E\right)\phi_- + \Delta\,\varphi_+ = 0. \quad (2.12)$$

Similar to the $\lambda = g$ case, solutions to these equations must be analytic in the whole complex plane if E belongs to the spectrum of $\tilde{H}_{dR}$. So we search for solutions of the form

$$\phi_+(z) = \sum_{n=0}^{\infty} \mathcal{S}_n^+(E)\, z^n, \qquad \phi_-(z) = \sum_{n=0}^{\infty} \mathcal{S}_n^-(E)\, z^n, \quad (2.13)$$

which are the elements of the BH space.

Substituting (2.13) into (2.12), we obtain

$$\mathcal{S}_n^- = \frac{\Delta}{E - n\omega + \delta + \frac{g^2}{\omega}}\,\mathcal{S}_n^+. \quad (2.14)$$

$\mathcal{S}_n^-$ is not analytic in E but has simple poles at

$$E = n\omega - \delta - \frac{g^2}{\omega}, \quad n = 0, 1, \cdots. \quad (2.15)$$

which gives the other set of exact energies of the driven Rabi model.

The energies (2.9) and (2.15) appear for special values of model parameters and correspond to the exceptional solutions of the driven Rabi model. When (2.9) and (2.15) are satisfied, the infinite series expansions (2.7) and (2.13) truncate and reduce to polynomials in z but only if the system parameters satisfy certain constraints. This can be easily verified by following the procedure in [16] (and thus the driven Rabi model is quasi-exactly solvable). The majority part of the spectrum of the driven Rabi model is regular for which (2.9) and (2.15) are not satisfied. The regular spectrum of the model can be obtained as follows.

From (2.6) and (2.11), we obtain the 3-term recurrence relation for $\mathcal{R}_n^-$ and $\mathcal{S}_n^+$, respectively,

$$\begin{aligned}
&\mathcal{R}_1^- + X_0\,\mathcal{R}_0^- = 0, \\
&\mathcal{R}_{n+1}^- + X_n\,\mathcal{R}_n^- + Y_n\,\mathcal{R}_{n-1}^- = 0, \quad n \geq 1,
\end{aligned} \quad (2.16)$$

$$S_1^+ + U_0 S_0^+ = 0,$$
$$S_{n+1}^+ + U_n S_n^+ + V_n S_{n-1}^+ = 0, \quad n \geq 1, \qquad (2.17)$$

where

$$X_n = \frac{1}{2g(n+1)} \left[E - n\omega + \delta - \frac{3g^2}{\omega} + \frac{\Delta^2}{E - n\omega - \delta + \frac{g^2}{\omega}} \right],$$

$$U_n = \frac{1}{2g(n+1)} \left[E - n\omega - \delta - \frac{3g^2}{\omega} + \frac{\Delta^2}{E - n\omega + \delta + \frac{g^2}{\omega}} \right],$$

$$Y_n = V_n = \frac{1}{n+1}. \qquad (2.18)$$

The characteristic equation for the $n \geq 1$ part of both (2.16) and (2.17) is given by $t^2 - \frac{\omega}{2g} t = 0$, which gives two distinct roots $t_1 = 0$ and $t_2 = \frac{\omega}{2g}$.

Let $\mathcal{R}_{n,1}^-, \mathcal{R}_{n,2}^-$ and $\mathcal{S}_{n,1}^+, \mathcal{S}_{n,2}^+$ denote the two linearly independent solutions of the 2nd equation of (2.16) and (2.17), respectively. Applying the Poincaré-Perron theorem (i.e., Theorems 2.1 and 2.2 of [24]), these solutions satisfy ($\mathcal{Z} = \mathcal{R}^-$ or $\mathcal{S}^+$)

$$\lim_{n \to \infty} \frac{\mathcal{Z}_{n+1,r}}{\mathcal{Z}_{n,r}} = t_r, \quad r = 1, 2. \qquad (2.19)$$

Thus $\mathcal{R}_n^{min} \equiv \mathcal{R}_{n,1}^-$ and $\mathcal{S}_n^{min} \equiv \mathcal{S}_{n,1}^+$ are minimal solutions. It is not difficult to show that the infinite series expansions (2.7) and (2.13) with coefficients given respectively by the minimal solution $\mathcal{R}_n^{min}$ and $\mathcal{S}_n^{min}$ are entire functions, i.e., the elements of the BH space.

We now proceed to find energy eigenvalues E corresponding to the minimal solutions $\mathcal{R}_n^{min}$ and $\mathcal{S}_n^{min}$. We follow a procedure presented in [25] that uses the relationship between minimal solutions and infinite continued fractions [24]. A similar procedure was applied in [10] to analyze the Rabi model.

The ratio of successive elements of the minimal solution sequences $\mathcal{R}_n^{min}$ and $\mathcal{S}_n^{min}$ are expressible in terms of continued fractions,

$$T_n = \frac{\mathcal{R}_{n+1}^{min}}{\mathcal{R}_n^{min}} = -\frac{Y_{n+1}}{X_{n+1}-} \frac{Y_{n+2}}{X_{n+2}-} \frac{Y_{n+3}}{X_{n+3}-} \cdots, \qquad (2.20)$$

$$T_n' = \frac{\mathcal{S}_{n+1}^{min}}{\mathcal{S}_n^{min}} = -\frac{V_{n+1}}{U_{n+1}-} \frac{V_{n+2}}{U_{n+2}-} \frac{V_{n+3}}{U_{n+3}-} \cdots, \qquad (2.21)$$

which for $n = 0$ reduce to, respectively

$$T_0 = \frac{\mathcal{R}_1^{min}}{\mathcal{R}_0^{min}} = -\frac{Y_1}{X_1-}\frac{Y_2}{X_2-}\frac{Y_3}{X_3-}\ldots. \quad (2.22)$$

$$T_0' = \frac{\mathcal{S}_1^{min}}{\mathcal{S}_0^{min}} = -\frac{V_1}{U_1-}\frac{V_2}{U_2-}\frac{V_3}{U_3-}\ldots. \quad (2.23)$$

Note that the ratios $T_0 = \frac{\mathcal{R}_1^{min}}{\mathcal{R}_0^{min}}$ and $T_0' = \frac{\mathcal{S}_1^{min}}{\mathcal{S}_0^{min}}$ involve $\mathcal{R}_0^{min}$ and $\mathcal{S}_0^{min}$, although the above continued fraction expressions are obtained from the 2nd equation of (2.16) and (2.17) i.e., those recurrence relations for $n \geq 1$. However, for the single-ended sequences appearing in the infinite series expansions (2.7) and (2.13), the first equation (i.e., the $n = 0$ part) of the recurrences (2.16) and (2.17) requires that

$$T_0 = -X_0 = -\frac{1}{2g}\left(E + \delta - \frac{3g^2}{\omega} + \frac{\Delta^2}{E - \delta + \frac{g^2}{\omega}}\right), \quad (2.24)$$

$$T_0' = -U_0 = -\frac{1}{2g}\left(E - \delta - \frac{3g^2}{\omega} + \frac{\Delta^2}{E + \delta + \frac{g^2}{\omega}}\right). \quad (2.25)$$

In general, (2.22), (2.24), (2.23), and (2.25) can not be satisfied at the same time for arbitrary values of the recurrence coefficients X_n, Y_n, U_n, and V_n. Physical meaningful solutions are those that are elements of the BH space [7]. They can be obtained if E can be adjusted so that equations (2.22), (2.24), (2.23) and (2.25) are all satisfied. Equating the right hand sides of (2.22), (2.23) and (2.24), (2.25), respectively, we obtain two transcendental functions $Q(E) = T_0 + X_0$ and $P(E) = T_0' + U_0$ for the spectrum E, where T_0, T_0' are the continued fractions in (2.22) and (2.23), while X_0, U_0 are the right-hand sides of (2.24) and (2.25). Then the zeros of $Q(E)$ and $P(E)$ correspond to the points in the parameter space where the conditions (2.24) and (2.25) are satisfied, i.e., the regular energies of the driven Rabi model are given by the zeros of the transcendental functions. The transcendental eigenvalue equations $Q(E) = 0$ and $P(E) = 0$ may be solved for E by standard root-search techniques (see e.g., [25, 26] and references therein). Only for the denumerable infinite values of E which are the roots of $Q(E) = 0$ and $P(E) = 0$, do we get entire solutions of the differential equations (2.5), (2.6), (2.11) and (2.12).

3. DRIVEN QUANTUM RABI MODEL

The Hamiltonian of the 2-photon Rabi model reads

$$H_{2p} = \omega b^\dagger b + \Delta \sigma_z + g\sigma_x \left[(b^\dagger)^2 + b^2\right]. \tag{3.1}$$

Let us make the canonical Bogoliubov transformation from $b, b^\dagger$ to squeezed bosons $a, a^\dagger$ [27],

$$b = \frac{a + \tau a^\dagger}{\sqrt{1 - \tau^2}}, \qquad b^\dagger = \frac{\tau a + a^\dagger}{\sqrt{1 - \tau^2}}, \tag{3.2}$$

where $|\tau| < 1$ is a real parameter. In terms of the squeezed bosons, the Hamiltonian (3.1) takes the form

$$\begin{aligned}\tilde{H}_{2p} &= \Delta \sigma_z + \frac{1}{1-\tau^2}\left[(\omega\tau + g\sigma_x(1+\tau^2))\left((a^\dagger)^2 + a^2\right)\right.\\ &\quad \left. + (\omega(1+\tau^2) + 4g\tau\sigma_x)\,a^\dagger a + \omega\tau^2 + 2g\tau\sigma_x\right].\end{aligned} \tag{3.3}$$

Introduce the operators $K_\pm, K_0$

$$K_+ = \frac{1}{2}(a^\dagger)^2, \quad K_- = \frac{1}{2}a^2, \quad K_0 = \frac{1}{2}\left(a^\dagger a + \frac{1}{2}\right). \tag{3.4}$$

Then (3.3) becomes

$$\begin{aligned}\tilde{H}_{2p} &= \Delta\sigma_z + \frac{1}{1-\tau^2}\Big[2\left(\omega\tau + g\sigma_x(1+\tau^2)\right)(K_+ + K_-) \\ &\quad + 2\left(\omega(1+\tau^2) + 4g\tau\sigma_x\right)K_0\Big] - \frac{1}{2}\omega.\end{aligned} \tag{3.5}$$

The operators $K_\pm, K_0$ form the $su(1,1)$ Lie algebra. Its quadratic Casimir, $C = K_+ K_- - K_0(K_0 - 1)$, takes the particular values $C = \frac{3}{16}$ in the representation (3.4). This is the well-known infinite-dimensional unitary irreducible representation $\mathcal{D}^+(q)$ of $su(1,1)$ with $q = \frac{1}{4}, \frac{3}{4}$. Thus the Fock-Hilbert space decomposes into the direct sum of two subspaces $\mathcal{H}^q$ labeled by $q = 1/4, 3/4$.

In the same way as the differential realization of boson operators in a Hilbert space of entire functions, we can represent [15, 16] the generators $K_\pm, K_0$ (3.4) as single-variable differential operators in Bargmann space $\mathcal{B}_q$ with basis vectors given by the monomials $\left\{z^n/\sqrt{[2(n+q-1/k^2)]!}\right\}$,

$$K_0 = z\frac{d}{dz} + q, \quad K_+ = \frac{z}{2}, \quad K_- = 2z\frac{d^2}{dz^2} + 4q\frac{d}{dz}. \qquad (3.6)$$

The Bargmann space $\mathcal{B}_q$ is the Hilbert space of entire functions on the complex plane if the inner product

$$\langle f|g\rangle_q = \int \overline{f(z)}\, g(z)\, d\mu_q(z) \qquad (3.7)$$

is finite for an appropriate measure $d\mu_q(z)$. Taking the measure to be [19]

$$d\mu_q(z) = \frac{1}{\pi}|z|^{2(q-3/4)} e^{-|z|}\, dx\, dy, \qquad (3.8)$$

then we can show by means of the formula $\Gamma(s) = \int_0^\infty \xi^{s-1} e^{-\xi} d\xi$ for $\mathrm{Re}(s) > 0$,

$$\langle z^m|z^n\rangle = [2(n+q-1/4)]!\, \delta_{mn}. \qquad (3.9)$$

Thus the monomials $\left\{z^n/\sqrt{[2(n+q-1/k^2)]!}\right\}$ form an orthonormal basis of the BH space $\mathcal{B}_q$. Note in passing that the standard measure $\frac{1}{\pi} e^{-|z|^2}\, dx\, dy$ is no longer appropriate here. It is now not difficult to see that if $f(z) = \sum_{n=0}^\infty c_n z^n$ then

$$\|f\|_q^2 = \sum_{n=0}^\infty |c_n|^2\, [2(n+q-1/4)]! \qquad (3.10)$$

and $f(z)$ is entire function in the BH space $\mathcal{B}_q$ if the sum on the right hand side converges.

Using this differential realization, working in a representation defined by σ_x diagonal and choosing τ to be the root of $\omega\tau + g(1+\tau^2) = 0$ so that it is real and obeys $|\tau| < 1$, i.e.,

$$\tau = -\frac{\omega}{2g}(1-\Omega), \quad \Omega = \sqrt{1 - \frac{4g^2}{\omega^2}}, \qquad (3.11)$$

where $\left|\frac{2g}{\omega}\right| < 1$, then the transformed 2-photon Rabi Hamiltonian becomes the matrix differential operator

$$\tilde{H}_{2p} = \begin{pmatrix} 2\omega\Omega\left(z\frac{d}{dz}+q\right)-\frac{1}{2}\omega & \Delta \\ \Delta & -\frac{8g}{\Omega}z\frac{d^2}{dz^2}+\frac{2}{\Omega}\left[\omega(2-\Omega^2)z-8gq\right]\frac{d}{dz} \\ & -\frac{2g}{\Omega}z+\frac{2\omega(2-\Omega^2)q}{\Omega}-\frac{1}{2}\omega \end{pmatrix}. \quad (3.12)$$

In terms of two-component wavefuntion $\psi(z) = (\psi_+(z), \psi_-(z))^T$, the time-independent Schrödinger equation, $\tilde{H}_{2p}\psi(z) = E\psi(z)$, yields a system of coupled differential equations,

$$\left[2\omega\Omega\left(z\frac{d}{dz}+q\right)-\frac{1}{2}\omega-E\right]\psi_+ + \Delta\psi_- = 0, \quad (3.13)$$

$$\left[8gz\frac{d^2}{dz^2}+\left(-2\omega(2-\Omega^2)z+16gq\right)\frac{d}{dz}\right.$$
$$\left.+2gz-2\omega(2-\Omega^2)q+(\frac{1}{2}\omega+E)\Omega\right]\psi_- - \Omega\Delta\psi_+ = 0 \quad (3.14)$$

This is a system of differential equations of Fuchsian type. Solutions to these equations must be analytic in the whole complex plane if E belongs to the spectrum of $\tilde{H}_{2p}$. So we are seeking solutions of the form

$$\psi_+(z) = \sum_{n=0}^{\infty} \mathcal{K}_n^+(E)\, z^n, \quad \psi_-(z) = \sum_{n=0}^{\infty} \mathcal{K}_n^-(E)\, z^n, \quad (3.15)$$

which converge in the entire complex plane, i.e., solutions which are entire.

Substituting (3.15) into (3.13), we obtain

$$\mathcal{K}_n^+ = \frac{\Delta}{E+\frac{1}{2}\omega-(2n+2q)\omega\Omega}\mathcal{K}_n^-. \quad (3.16)$$

So $\mathcal{K}_n^+$ is not analytic in E but has simple poles at

$$E = -\frac{1}{2}\omega + (2n+2q)\omega\Omega, \quad n=0,1,\cdots. \quad (3.17)$$

The energies (3.17) appear for special values of model parameters [16, 27] and correspond to the exceptional solutions of the 2-photon Rabi model. If (3.17) is satisfied, the infinite series expansions (3.15) truncate and reduce to polynomials in z but only if the system parameters satisfy certain constraints [16]. The majority part of the spectrum of the 2-photon Rabi model is regular for which (3.17) is not satisfied. The regular spectrum of the model is given by the zeros of the transcendental function $F(E)$ obtained below. Thus similar to the Rabi case, the spectrum of the 2-photon Rabi model consists of two parts, the regular and the exceptional spectrum.

From (3.14), we obtain the 3-step recurrence relation for $\mathcal{K}_n^-$ [28],

$$\mathcal{K}_1^- + A_0 \mathcal{K}_0^- = 0,$$
$$\mathcal{K}_{n+1}^- + A_n \mathcal{K}_n^- + B_n \mathcal{K}_{n-1}^- = 0, \quad n \geq 1, \tag{3.18}$$

where

$$A_n = \frac{1}{8g(n+1)(n+2q)}\left[-(2n+2q)\omega(2-\Omega^2)\right.$$
$$\left. + \left(E + \frac{1}{2}\omega - \frac{\Delta^2}{E + \frac{1}{2}\omega - (2n+2q)\omega\Omega}\right)\Omega\right],$$
$$B_n = \frac{1}{4(n+1)(n+2q)}. \tag{3.19}$$

The coefficients A_n, B_n have the behavior as $n \to \infty$

$$A_n \sim a\, n^\alpha, \quad B_n \sim b\, n^\beta, \tag{3.20}$$

with

$$a = -\frac{\omega}{4g}(2-\Omega^2), \quad \alpha = -1, \quad b = \frac{1}{4}, \quad \beta = -2. \tag{3.21}$$

Thus the asymptotic structure of solutions to the 2nd equation of (3.18) depends on the Newton-Puiseux diagram formed with the points $P_0(0,0), P_1(1,-1), P_2(2,-2)$ [24]. Let γ be the slope of $\overline{P_0 P_1}$ and δ the slope of $\overline{P_1 P_2}$ so that $\gamma = \alpha$ and $\delta = \beta - \alpha$. Then we have $\gamma = \delta = \alpha$. The characteristic equation of the $n \geq 1$ part of (3.18) reads $t^2 + at + b = 0$ with a, b given in (3.21). It has two roots $t_1 = \frac{\omega}{4g}, t_2 = \frac{g}{\omega}$. Remembering the condition $\left|\frac{2g}{\omega}\right| < 1$, we have $|t_2| < |t_1|$. Applying the Perron-Kreuser theorem (i.e., Theorem 2.3 of [24]), we conclude that the two linearly independent solutions $\mathcal{K}_{n,1}^-$ and $\mathcal{K}_{n,2}^-$ of the $n \geq 1$ part (i.e., the truly 3-term part) of (3.18) satisfy

$$\lim_{n \to \infty} \frac{\mathcal{K}_{n+1,r}^-}{\mathcal{K}_{n,r}^-} \sim t_r\, n^{-1}, \quad r = 1, 2. \tag{3.22}$$

So $\mathcal{K}_{n,2}^-$ is a minimal solution and $\mathcal{K}_{n,1}^-$ is a dominant one. By (3.10), we can see that the infinite power series in (3.15) with expansion coefficients $\mathcal{K}_{n,r}^-$ is entire if the sum

$$\sum_{n=0}^{\infty} |\mathcal{K}_{n,r}^-|^2 [2(n+q-1/4)]! \qquad (3.23)$$

converges. Using the asymptotic form (3.22), we get

$$\lim_{n\to\infty} \frac{\left|\mathcal{K}_{n+1,r}^-\right|^2 [2(n+1+q-1/4)]!}{\left|\mathcal{K}_{n,r}^-\right|^2 [2(n+q-1/4)]!} = 4|t_r|^2, \qquad (3.24)$$

which is less than 1 for $r = 2$ and greater than 1 for $r = 1$. Thus by the ratio test, the sum (3.23) converges for the minimal solution $\mathcal{K}_n^{min} \equiv \mathcal{K}_{n,2}^-$ and diverges for the dominant solution $\mathcal{K}_{n,1}^-$. It follows that the infinite power series expansions $\psi_\pm^{min}(z)$, obtained by substituting $\mathcal{K}_n^{min}$ for the $\mathcal{K}_n^-$'s in (3.16) and (3.15), converge in the whole complex plane, i.e., they are entire.

By the Pincherle theorem (i.e., Theorem 1.1 of [24]), the ratio of successive elements of the minimal solution sequence $\mathcal{K}_n^{min}$ is expressible as continued fractions,

$$R_n = \frac{\mathcal{K}_{n+1}^{min}}{\mathcal{K}_n^{min}} = -\frac{B_{n+1}}{A_{n+1}-}\frac{B_{n+2}}{A_{n+2}-}\frac{B_{n+3}}{A_{n+3}-}\cdots, \qquad (3.25)$$

which for $n = 0$ gives

$$R_0 = \frac{\mathcal{K}_1^{min}}{\mathcal{K}_0^{min}} = -\frac{B_1}{A_1-}\frac{B_2}{A_2-}\frac{B_3}{A_3-}\cdots. \qquad (3.26)$$

Note that the ratio $R_0 = \frac{\mathcal{K}_1^{min}}{\mathcal{K}_0^{min}}$ involves $\mathcal{K}_0^{min}$, although the above continued fraction expression is obtained from the 2nd equation of (3.18), i.e., the recurrence (3.18) for $n \geq 1$. On the other hand, the ratio $R_0 = \frac{\mathcal{K}_1^{min}}{\mathcal{K}_0^{min}}$ of the first two terms of a minimal solution is unambiguously fixed by the first equation of the recurrence (3.18), namely,

$$R_0 = -A_0 = \frac{1}{16gq}\left[2q\omega(2-\Omega^2) - \left(E + \frac{1}{2}\omega - \frac{\Delta^2}{E+\frac{1}{2}\omega - 2q\omega\Omega}\right)\Omega\right]. \qquad (3.27)$$

In general, the R_0 computed from the continued fraction (3.26) can not be the same as that from (3.27) for arbitrary values of recurrence coefficients A_n and

B_n. Following similar discussions in the last section for the regular energy spectrum of the driven Rabi model, entire analytic function solutions which are elements of the BH space $\mathcal{B}_q$ require E be the roots of the transcendental equation $F(E) = R_0 + A_0 = 0$ with R_0 given by the continued fraction in (3.26). In other words, $F(E) = 0$ is the eigenvalue equation of the 2-photon Rabi model. Only for the denumerable infinite values of E which are the roots of $F(E) = 0$, do we get entire analytic function solutions of the differential equations (3.13) and (3.14) which are normalizable with respect to the BH norm (3.7).

4. Two-Mode Quantum Rabi Model

We consider the Hamiltonian of the two-mode Rabi model introduced in [16]

$$H_{2m} = \omega(b_1^\dagger b_1 + b_2^\dagger b_2) + \Delta\sigma_z + g\,\sigma_x(b_1^\dagger b_2^\dagger + b_1 b_2), \tag{4.1}$$

where we assume that the boson modes are degenerate with the same frequency ω. Introduce the two-mode Bogoliubov transformation,

$$b_1 = \frac{a_1 + \sigma a_2^\dagger}{\sqrt{1-\sigma^2}}, \quad b_1^\dagger = \frac{\sigma a_2 + a_1^\dagger}{\sqrt{1-\sigma^2}}, \quad b_2 = \frac{a_2 + \sigma a_1^\dagger}{\sqrt{1-\sigma^2}}, \quad b_2^\dagger = \frac{\sigma a_1 + a_2^\dagger}{\sqrt{1-\sigma^2}}. \tag{4.2}$$

Here $|\sigma| < 1$ is a real parameter and $a_1, a_2, a_1^\dagger, a_2^\dagger$ are squeezed bosons satisfying the canonical commutation relations $[a_i, a_i^\dagger] = 1$, $[a_i, a_j] = [a_i, a_j^\dagger] = [a_i^\dagger, a_j^\dagger] = 0$, $i, j = 1, 2$. In terms of the 2-mode squeezed bosons, the Hamiltonian (4.1) has the form

$$\begin{aligned}\tilde{H}_{2m} &= \frac{1}{1-\sigma^2}\Big[(2\omega\sigma + g\sigma_x(1+\sigma^2))\,(a_1^\dagger a_2^\dagger + a_1 a_2) \\ &\quad + (\omega(1+\sigma^2) + 2g\sigma\,\sigma_x)(a_1^\dagger a_1 + a_2^\dagger a_2) + 2\omega\sigma^2 + 2g\sigma\,\sigma_x\Big] + \Delta\sigma_z.\end{aligned} \tag{4.3}$$

Introduce the operators $K_\pm, K_0$

$$K_+ = a_1^\dagger a_2^\dagger, \quad K_- = a_1 a_2, \quad K_0 = \frac{1}{2}(a_1^\dagger a_1 + a_2^\dagger a_2 + 1). \tag{4.4}$$

Then (4.3) becomes

$$\begin{aligned}\tilde{H}_{2m} &= \Delta\sigma_z + \frac{1}{1-\sigma^2}\Big[(2\omega\sigma + g\sigma_x(1+\sigma^2))\,(K_+ + K_-) \\ &\quad + 2\left(\omega(1+\sigma^2) + 2g\sigma\sigma_x\right)K_0\Big] - \omega.\end{aligned} \tag{4.5}$$

The operators $K_\pm, K_0$ form the $su(1,1)$ Lie algebra. Its quadratic Casimir, $C = K_+K_- - K_0(K_0 - 1)$, takes the particular values $C = \kappa(1-\kappa)$ in the representation (4.4), where $\kappa = 1/2, 1, 3/2, \cdots$. This is the well-known infinite-dimensional unitary irreducible representation of $su(1,1)$ known as the positive discrete series $\mathcal{D}^+(\kappa)$. Thus the Fock-Hilbert space decomposes into the direct sum of infinite subspaces $\mathcal{H}^\kappa$ labeled by $\kappa = 1/2, 1, 3/2, \cdots$.

Similar to the 2-photon Rabi case, we can represent [15, 16] the generators $K_\pm, K_0$ (4.4) as single-variable differential operators in z in Bargmann space $\mathcal{B}_\kappa$ with basis vectors given by the monomials $\left\{z^n/\sqrt{n!(n+2\kappa-1)!}\right\}$,

$$K_0 = z\frac{d}{dz} + \kappa, \quad K_+ = z, \quad K_- = z\frac{d^2}{dz^2} + 2\kappa\frac{d}{dz}, \tag{4.6}$$

where $\kappa = 1/2, 1, 3/2, \cdots$. The Bargmann space $\mathcal{B}_\kappa$ is the Hilbert space of entire functions in z if the inner product

$$\langle f|g\rangle_\kappa = \int \overline{f(z)}\,g(z)\,d\mu_\kappa(z) \tag{4.7}$$

is finite for an appropriate measure $d_\kappa\mu(z)$. It can be shown [29] that if we choose

$$d\mu_\kappa(z) = \frac{4}{\pi}|z|^{2\kappa-1}K_{1/2-\kappa}(2|z|)\,dx\,dy, \tag{4.8}$$

where $K_\nu(z)$ is the modified Bessel function of the third kind which has the Mellin transform $\int_0^\infty 2\xi^{\alpha+\beta}K_{\alpha-\beta}(2\xi^{1/2})\xi^{s-1}d\xi = \Gamma(s+2\alpha)\Gamma(s+2\beta)$, then

$$\langle z^m|z^n\rangle_\kappa = n!\,(n+2\kappa-1)!\,\delta_{mn}. \tag{4.9}$$

Thus the monomials $\left\{z^n/\sqrt{n!(n+2\kappa-1)!}\right\}$ form an orthonormal basis of the BH space $\mathcal{B}_\kappa$. It is now not difficult to see that if $f(z) = \sum_{n=0}^\infty c_n z^n$ then

$$\|f\|_\kappa^2 = \sum_{n=0}^\infty |c_n|^2\,n!\,(n+2\kappa-1)! \tag{4.10}$$

and $f(z)$ is entire function belonging to $\mathcal{B}_\kappa$ if the sum on the right hand side converges.

Using this differential realization, working in a representation defined by σ_x diagonal and choosing σ to be the root of $2\omega\sigma + g(1+\sigma^2) = 0$ so that it is real and satisfies $|\sigma| < 1$, i.e.,

$$\sigma = -\frac{\omega}{g}(1-\Lambda), \quad \Lambda = \sqrt{1-\frac{g^2}{\omega^2}}, \tag{4.11}$$

where $\left|\frac{g}{\omega}\right| < 1$, then the transformed Hamiltonian (4.5) becomes a matrix differential operator

$$\tilde{H}_{2m} = \begin{pmatrix} 2\omega\Lambda\left(z\frac{d}{dz}+\kappa\right) - \omega & \Delta \\ \Delta & -\frac{2g}{\Lambda}z\frac{d^2}{dz^2} + \frac{2}{\Lambda}\left[\omega(2-\Lambda^2)z - 2g\kappa\right]\frac{d}{dz} \\ & -\frac{2g}{\Lambda}z + \frac{2\omega(2-\Lambda^2)\kappa}{\Lambda} - \omega \end{pmatrix}. \tag{4.12}$$

In terms of two-component wavefuntion $\phi(z) = (\phi_+(z), \phi_-(z))^T$, the time-independent Schrödinger equation, $\tilde{H}_{2m}\phi(z) = E\phi(z)$, yields a system of coupled differential equations,

$$\left[2\omega\Lambda\left(z\frac{d}{dz}+\kappa\right) - \omega - E\right]\phi_+ + \Delta\phi_- = 0, \tag{4.13}$$

$$\left[2gz\frac{d^2}{dz^2} + \left(-2\omega(2-\Lambda^2)z + 4g\kappa\right)\frac{d}{dz} + 2gz - 2\omega(2-\Lambda^2)\kappa + (E+\omega)\Lambda\right]\phi_- - \Lambda\Delta\,\phi_+ = 0. \tag{4.14}$$

This is a system of differential equations of Fuchsian type. Solutions to these equations must be analytic in the whole complex plane if E belongs to the spectrum of $\tilde{H}_{2m}$. Similar to the 2-photon Rabi case, we seek solutions of the form

$$\phi_+(z) = \sum_{n=0}^{\infty} \mathcal{Q}_n^+(E)\,z^n, \quad \phi_-(z) = \sum_{n=0}^{\infty} \mathcal{Q}_n^-(E)\,z^n, \tag{4.15}$$

which converge in the entire complex plane.

Substituting (4.15) into (4.13), we obtain

$$\mathcal{Q}_n^+ = \frac{\Delta}{E+\omega - (2n+2\kappa)\omega\Lambda}\mathcal{Q}_n^-. \tag{4.16}$$

So $\mathcal{Q}_n^+$ is not analytic in E but has simple poles at

$$E = -\omega + (2n+2\kappa)\omega\Lambda, \quad n = 0, 1, \cdots. \tag{4.17}$$

The energies (4.17) appear for special values of model parameters [16] and correspond to the exceptional solutions of the two-mode Rabi model. If (4.17) is satisfied, the infinite series expansions (4.15) truncate and reduce to polynomials in z but only if the model parameters obey certain constraints [16]. The majority part of the spectrum of the two-mode Rabi model is regular spectrum which does not have the form (4.17). The regular spectrum of the model is given by the zeros of the transcendental function $G(E)$ obtained below. Thus again, the spectrum of the two-mode Rabi model consists of two parts, the regular and the exceptional spectrum.

From (4.14), we obtain the 3-step recurrence relation for $\mathcal{Q}_n^-$ [28],

$$\mathcal{Q}_1^- + C_0 \mathcal{Q}_0^- = 0,$$
$$\mathcal{Q}_{n+1}^- + C_n \mathcal{Q}_n^- + D_n \mathcal{Q}_{n-1}^- = 0, \quad n \geq 1, \quad (4.18)$$

where

$$C_n = \frac{1}{2g(n+1)(n+2\kappa)} \left[-(2n+2\kappa)\omega(2-\Lambda^2) \right.$$
$$\left. + \left(E + \omega - \frac{\Delta^2}{E+\omega-(2n+2\kappa)\omega\Lambda} \right)\Lambda \right],$$
$$D_n = \frac{1}{(n+1)(n+2\kappa)}. \quad (4.19)$$

The coefficients C_n, D_n have the behavior as $n \to \infty$

$$C_n \sim c\, n^\mu, \quad D_n \sim d\, n^\rho, \quad (4.20)$$

with

$$c = -\frac{\omega}{g}(2-\Lambda^2), \quad \mu = -1, \quad d = 1, \quad \rho = -2. \quad (4.21)$$

By analysis similar to the 2-photon case, we see that the two linearly independent solutions $\mathcal{Q}_{n,1}^-$ and $\mathcal{Q}_{n,2}^-$ of the $n \geq 1$ part of the recurrence (3.18) obey

$$\lim_{n\to\infty} \frac{\mathcal{Q}_{n+1,r}^-}{\mathcal{Q}_{n,r}^-} \sim t_r\, n^{-1}, \quad r = 1, 2, \quad (4.22)$$

where $t_1 = \frac{\omega}{g}$, $t_2 = \frac{g}{\omega}$ and $|t_2| < |t_1|$ (from the condition $\left|\frac{g}{\omega}\right| < 1$). Thus $\mathcal{Q}_{n,2}^-$ is a minimal solution and $\mathcal{Q}_{n,1}^-$ is a dominant one. Using (4.10) and by similar analysis to the 2-photon case, we can conclude that the infinite power series expansions $\phi_{\pm}^{min}(z)$ generated by substituting the minimal solution $\mathcal{Q}_n^{min} \equiv \mathcal{Q}_{n,2}^{-}$ for the $\mathcal{Q}_n^{-}$'s in (4.16) and (4.15), converge in the whole complex plane.

From the 2nd equation of (4.18), the ratio of successive elements of the minimal solution $\mathcal{Q}_n^{min}$ can be expressed as continued fractions,

$$S_n = \frac{\mathcal{Q}_{n+1}^{min}}{\mathcal{Q}_n^{min}} = -\frac{D_{n+1}}{C_{n+1}-}\frac{D_{n+2}}{C_{n+2}-}\frac{D_{n+3}}{C_{n+3}-}\cdots, \qquad (4.23)$$

which for $n = 0$ reduces to

$$S_0 = \frac{\mathcal{Q}_1^{min}}{\mathcal{Q}_0^{min}} = -\frac{D_1}{C_1-}\frac{D_2}{C_2-}\frac{D_3}{C_3-}\cdots. \qquad (4.24)$$

On the other hand, the ratio $S_0 = \frac{\mathcal{Q}_1^{min}}{\mathcal{Q}_0^{min}}$ of the first two terms of a minimal solution is unambiguously fixed by the $n = 0$ part of the recurrence (4.18), that is,

$$S_0 = -C_0 = \frac{1}{4g\kappa}\left[2\kappa\omega(2-\Lambda^2) - \left(E+\omega - \frac{\Delta^2}{E+\omega-2\kappa\omega\Lambda}\right)\Lambda\right]. \qquad (4.25)$$

Then similar to the 2-photon Rabi case, entire power series solutions which are elements of the BH space $\mathcal{B}_\kappa$ require that E can be adjusted so that equations (4.24) and (4.25) are both satisfied. This yields an implicit continued fraction equation for the regular spectrum E. That is, the regular energies E of the two-mode Rabi model are determined by the zeros of the transcendental function $G(E) = S_0 + C_0$ with S_0 and C_0 given by (4.24) and (4.25), respectively. Only for the denumerable infinite values of E which are the roots of $G(E) = 0$, do we get entire analytic function solutions of the differential equations (4.13) and (4.14) which are normalizable with respect to the BH norm (4.7).

ACKNOWLEDGMENTS

This work was partially supported by the Australian Research Council Discovery-Projects grant DP190101529.

CONCLUSION

We have presented the analytic representations of solutions to the driven, 2-photon, and two-mode quantum Rabi models. The regular eigenvalues of the models are given by the zeros of the transcendental functions, which have been analytically found by applying the Bogoliubov transformations and the new BH spaces.

This work should be of general interest due to its focus on the non-trivial generalizations of the popular quantum Rabi model, but also be intriguing to those interested in analytic solutions of Fuchsian differential equations in physics.

REFERENCES

[1] Vedral V. (2006), *Modern foundations of quantum optics*. Imperial College Press, London.

[2] Khitrova G., Gibbs H. M., Kira M., Koch S. W. and Scherer A. (2006), Vacuum Rabi splitting in semiconductor. *Nature Physics*, 2, 81.

[3] Thanopulos I., Paspalakis E. and Kis Z. (2004), Laser-driven coherent manipulation of molecular chirality. *Chemical Physics Letters*, 390, 228.

[4] Englund D., Faraon A., Fushman I., Stoltz N., Petroff P. and Vuckovic J. (2007), Controlling cavity reflectivity with a single quantum dot. *Nature*, 450, 857.

[5] Niemczyk T., Deppe F., Huebl H., Menzel E. P., Hocke F., Schward M. J., Garcia-Ripoll J. J., Zueco D., Hümmer T., Solano E., Marx A. and Gross, R. (2010), Circuit quantum elctrodynamics in ultrastrong-coupling regime. *Nature Physics*, 6, 772.

[6] Bargmann V. (1961), On a Hilbert space of analytic functions and associated integral transform part I. *Communications on Pure and Applied Mathematics*, 14, 187.

[7] Schweber S. (1967), On the application of Bargmann Hilbert spaces to dynamical problems. *Annals of Physics*, 41, 205.

[8] Braak D. (2011), On the integrability of the Rabi model. *Physical Review Letter*, 107, 100401.

[9] Solano E. (2011), Viepoint: The dialogue between quantum light and matter. *Physics*, 4, 68.

[10] Moroz A. (2012), On the spectrum of class of quantum models. *Europhysics Letters*, 100, 60010.

[11] Maciejewski A. J., Przybylska M. and Stachowiak T. (2012), How to calculate spectra of Rabi and related models. arXiv:1210.1130 [math-ph].

[12] Travénec I. (2012), Solvability of the two-photon Rabi Hamiltonian. *Physical Review A*, 85, 043805.

[13] Chen Q. H., Wang C., He S., Liu T. and Wang K. L. (2012), Exact solvability of the quantum Rabi model using Bogoliubov operators. *Physical Review A*, 86, 023822.

[14] Moroz A. (2013), On solvability and integrability of the Rabi model. *Annals of Physics*, 338, 319.

[15] Zhang Y.-Z. (2013), Solving two-mode squeezed harmonic oscillator and kth-order harmonic generation in Bargmann-Hilbert spaces. *Journal of Physics A: Mathematical and Theoretical*, 46, 455302.

[16] Zhang Y.-Z. (2013), On the solvability of the quantum Rabi model and its 2-photon and two-mode generalizations. *Journal of Mathematical Physics*, 54, 102104.

[17] Zhong H., Xie Q., Batchelor M. T. and Lee C. (2013), Analytical eigenstates for the quantum Rabi model. *Journal of Physics A: Mathematical and Theoretical*, 46, 415302.

[18] Moroz A. (2014), A hidden analytic structure of the Rabi model. *Annals of Physics*, 340, 252.

[19] Zhang Y.-Z. (2017), On the 2-mode and k-photon quantum Rabi models. *Reviews in Mathematical Physics*, 29, 1750013.

[20] Larson J. (2013), Integrability vs quantum thermalization. *Journal of Physics B: Atomic, Molecular and Optical Physics*, 46, 224016.

[21] Brune M., Raimond J. M., Goy P., Davidovich L. and Haroche S. (1987), Realization of a two-photon mass oscillator. *Physical Review Letter*, 59, 1899.

[22] del Valle E., Zippilli S., Laussy F. P., Gonzalez-Tudela A., Morigi G. and Tejedor C. (2012), Tow-photon lasing by a single quantum dot in a high-Q microcavity. *Physical Review B*, 81, 035302.

[23] Ota Y., Iwamoto S., Kumagai N. and Arakawa Y. (2011), Spontaneous two photon emission from a single quantum dot. *Physical Review Letter*, 107, 233602.

[24] Gautschi W. (1967), Computational aspects of three-term recurrence relations. *SIAM Review*, 9, 24.

[25] Leaver E. W. (1986), Solutions to a generalized spheroidal wave equation: Teukolsky's equations in general relativity, and the two-center problem in molecular quantum mechanics. *Journal of Mathematical Physics*, 27, 1238.

[26] Liu J. W. (1992), Analytical solutions to the generalized spheroidal wave equation and the Green's function of one-electron diatomic molecules. *Journal of Mathematical Physics*, 33, 4026.

[27] Emary C. and Bishop R. F. (2002), Exact isolated solutions for the two-photon Rabi Hamiltonian. *Journal of Physics A: Mathematical and Theoretical*, 35, 8231.

[28] Zhang Y.-Z. (2019), Quasi-exactly solvable systems. Lectures given at the 9th NSFC Summer School in Theoretical Physics, Xi'an, 2016. In "Integrable Model Methods and Applications", edited by W.-L. Yang, Z.-Y. Yang and T. Yang, Science Press, Beijing, China.

[29] Barut A. O. and Girardello L. (1971), New "coherent" states associated with non-compact groups. *Communications in Mathematical Physics*, 21, 41.

In: Hilbert Spaces: Properties and Applications ISBN: 978-1-53616-633-0
Editor: Le Bin Ho © 2020 Nova Science Publishers, Inc.

Chapter 6

HILBERT SPACE OF MODEL HAMILTONIANS

Medha Sharma[*]
Mater Dei School, New Delhi, India

Abstract

In this chapter, we show the method of generation of basis states and the formation of the Hamiltonian matrix (used in numerical techniques like exact diagonalization) of the Hubbard model which is the simplest model used in theoretical condensed matter physics mainly in the study of strongly correlated electron systems. This method of generation of basis states is applicable to other variants of the Hubbard model like the Anderson impurity model (used in dynamical mean-field theory), extended Hubbard model, etc.

Keywords: strongly correlated electron systems, Hubbard model, Hamiltonian matrix, exact diagonalization

1. INTRODUCTION

Hubbard model [1] was introduced to study correlation effects and ferromagnetism in narrow transition metals. This model is exactly solvable in one dimension [2].

In higher dimensions, the Hubbard model can be solved using numerical methods like quantum Monte Carlo [3], exact diagonalization [4–6]. After the

---
[*]Corresponding Author's E-mail: medhajamia@gmail.com.

discovery of high T_c superconductors [7], the interest in the Hubbard model was renewed as it turned out to be the simplest model to study the high-temperature superconductivity. In the infinite limit [8], the dynamical mean-field theory [9], that freezes the spatial fluctuations maps the Hubbard model onto Anderson impurity model [10] which can further be solved by numerical methods [11,12].

In this chapter, we show how to generate the basis states of the Hubbard model using binary numbers [13, 14], the action of fermionic operators on the basis states to form the Hamiltonian matrix. The Hamiltonian matrix can be further diagonalized to get the eigenvalues and eigenvectors.

2. MODEL SYSTEM

The Hamiltonian of a one-band Hubbard model is

$$H = -t \sum_{\langle i,j \rangle, \sigma} (c_{i\sigma}^\dagger c_{j\sigma} + c_{j\sigma}^\dagger c_{i\sigma}) + U \sum_i n_{i\uparrow} n_{i\downarrow}, \qquad (2.1)$$

where $c_{i\sigma}^\dagger$, $c_{i\sigma}$, and $n_{i\sigma} = c_{i\sigma}^\dagger c_{i\sigma}$ are fermion creation, annihilation, and number operators respectively at site i with spin $\sigma(\uparrow$ or $\downarrow)$. The angular bracket $\langle i, j \rangle$ gives the sum over nearest neighbors. U is the onsite repulsion and t is the hopping amplitude.

3. BASIS STATES

Each site has four possible options:

  i) $|0\rangle$: Empty

  ii) $|\uparrow\rangle$: Singly occupied with spin up electron

  iii) $|\downarrow\rangle$: Singly occupied with spin down electron

  iv) $|\uparrow\downarrow\rangle$: Doubly occupied with two opposite spin electrons.

If we consider a model of M sites, total possible basis states would be 4^M, which is quite a big number. Therefore, to reduce the size of the Hilbert space, we use various symmetries of the system, for example, translational invariance [15]. The most commonly used symmetries are S^z [z-component of spin

$(N_\uparrow - N_\downarrow)/2$] and number operator ($N = N_\uparrow + N_\downarrow$). If we employ both the symmetries, we end up with a set of basis states with a fixed number of spin up and a fixed number of spin down electrons (called a **sector**). If we work in a sector, the dimension of Hilbert space is considerably reduced. For example, consider a 4×4 square lattice with 16 sites, the total number of basis states would be $4^{16} = 4294967296$, but the number of basis states in the largest possible sector in the case of 16 sites, i.e., [($N_\uparrow = 8, N_\downarrow = 8$)], is $^{16}C_8 \times {}^{16}C_8 = 165636900$.

4. BIT REPRESENTATION

If we consider one particular spin at a time, a given site can either be occupied or unoccupied, thus giving rise to two states. These two states are mutually exclusive and exhaustive, thus could be perfectly represented by the binary numbers 0 and 1.

Suppose we consider a two-site Hubbard model, the spin up electrons can have the following possibilities:

$|00\rangle = 0$ (integral value)

$|0 \uparrow\rangle = 1$ (integral value)

$|\uparrow 0\rangle = 2$ (integral value)

$|\uparrow\uparrow\rangle = 3$ (integral value).

Similarly, the spin down electrons will also have the following possibilities:

$|00\rangle = 0$ (integral value)

$|0 \downarrow\rangle = 1$ (integral value)

$|\downarrow 0\rangle = 2$ (integral value)

$|\downarrow\downarrow\rangle = 3$ (integral value).

The conventional way of representing a basis state is combining both up and down basis states, denoted by $I = I_\uparrow + 2^M I_\downarrow$ [16], giving the following basis states:

$|00\rangle |00\rangle = 0$ (integral value)

$|0\uparrow\rangle|00\rangle = 1$ (integral value)

$|\uparrow 0\rangle|00\rangle = 2$ (integral value)

$|\uparrow\uparrow\rangle|00\rangle = 3$ (integral value)

$|00\rangle|0\downarrow\rangle = 4$ (integral value)

$|0\uparrow\rangle|0\downarrow\rangle = 5$ (integral value)

$|\uparrow 0\rangle|0\downarrow\rangle = 6$ (integral value)

$|\uparrow\uparrow\rangle|0\downarrow\rangle = 7$ (integral value)

$|00\rangle|\downarrow 0\rangle = 8$ (integral value)

$|0\uparrow\rangle|\downarrow 0\rangle = 9$ (integral value)

$|\uparrow 0\rangle|\downarrow 0\rangle = 10$ (integral value)

$|\uparrow\uparrow\rangle|\downarrow 0\rangle = 11$ (integral value)

$|00\rangle|\downarrow\downarrow\rangle = 12$ (integral value)

$|0\uparrow\rangle|\downarrow\downarrow\rangle = 13$ (integral value)

$|\uparrow 0\rangle|\downarrow\downarrow\rangle = 14$ (integral value)

$|\uparrow\uparrow\rangle|\downarrow\downarrow\rangle = 15$ (integral value).

Since in Hubbard model and other similar models, the spin up and spin down electrons do not mix with each other; we are able to deal with them separately. We just need to generate and work with one spin basis at a time [17].

5. Action of Operators on the Basis States

The operator $c_{j\sigma}$ annihilates the spin σ electron at site j and the operator $c_{i\sigma}^{\dagger}$ creates a spin σ electron at site i. While applying these operators, $c_{i\sigma}^{\dagger}c_{j\sigma}$, the phase change is also taken care of, due to the anti-commutation relation $\{c_i^{\dagger}c_j\} = \delta_{ij}$

of fermionic operators. The phase factor is actually $(-1)^n$, where n is the number of occupied sites between i and j.
For instance,

$$-tc_{1\uparrow}^\dagger c_{2\uparrow} |\uparrow 0\rangle |\downarrow 0\rangle \rightarrow -t |0 \uparrow\rangle |\downarrow 0\rangle, \qquad (5.1)$$

the spin up electron on the second site is annihilated and first site is created.

U takes into account the double occupancy of a site. For instance,

$$U n_{2\uparrow} n_{2\downarrow} |\uparrow 0\rangle |\downarrow 0\rangle \rightarrow U |\uparrow 0\rangle |\downarrow 0\rangle. \qquad (5.2)$$

6. AN EXAMPLE

The above steps are illustrated with the help of an example. Consider a sector($N_\uparrow = 3$, $N_\downarrow = 3$) of a 4-site Hubbard model. Let us generate the spin up basis states. The total number of spin up basis states is $^4C_3 = 4$.

The spin up basis states are:

$|0111\rangle = 7$ (integral value)

$|1011\rangle = 11$ (integral value)

$|1101\rangle = 13$ (integral value)

$|1110\rangle = 14$ (integral value).

Considering the nearest neighbor hopping and accounting the periodic boundary condition, we get:

$$-tc_{i\uparrow}^\dagger c_{j\uparrow} = -(tc_{2\uparrow}^\dagger c_{1\uparrow} + tc_{3\uparrow}^\dagger c_{2\uparrow} + tc_{4\uparrow}^\dagger c_{3\uparrow} + tc_{1\uparrow}^\dagger c_{4\uparrow} + tc_{1\uparrow}^\dagger c_{2\uparrow} + tc_{2\uparrow}^\dagger c_{3\uparrow} + tc_{3\uparrow}^\dagger c_{4\uparrow} + tc_{4\uparrow}^\dagger c_{1\uparrow}.$$

and their action on the basis states:

$$-(tc_{4\uparrow}^\dagger c_{3\uparrow} + tc_{4\uparrow}^\dagger c_{1\uparrow}) |0111\rangle \rightarrow (t|1011\rangle + t|1110\rangle)$$

$$-(tc_{3\uparrow}^\dagger c_{2\uparrow} + tc_{3\uparrow}^\dagger c_{4\uparrow}) |1011\rangle \rightarrow -(t|1101\rangle + t|1110\rangle)$$

$$-(tc_{2\uparrow}^\dagger c_{3\uparrow} + tc_{2\uparrow}^\dagger c_{1\uparrow}) |1101\rangle \rightarrow -(t|1011\rangle + t|1110\rangle$$

$$-(tc_{1\uparrow}^\dagger c_{4\uparrow} + tc_{1\uparrow}^\dagger c_{2\uparrow}) |1110\rangle \rightarrow -(t|0111\rangle + t|1101\rangle).$$

Due to the spin up hopping terms, we get the following Hamiltonian:

$$H_\uparrow = \begin{pmatrix} 0 & -t & 0 & -t \\ -t & 0 & -t & 0 \\ 0 & -t & 0 & -t \\ -t & 0 & -t & 0 \end{pmatrix}. \qquad (6.1)$$

Similarly, due to spin down hopping terms, we obtain the following Hamiltonian:

$$H_\downarrow = \begin{pmatrix} 0 & -t & 0 & -t \\ -t & 0 & -t & 0 \\ 0 & -t & 0 & -t \\ -t & 0 & -t & 0 \end{pmatrix}. \qquad (6.2)$$

The complete non-diagonal Hamiltonian matrix is obtained from $H_\uparrow$ and $H_\downarrow$ using the relation:

$$\mathcal{I}_\downarrow \otimes H_\uparrow \oplus H_\downarrow \otimes \mathcal{I}_\uparrow, \qquad (6.3)$$

where $\mathcal{I}_\downarrow$ and $\mathcal{I}_\uparrow$ are the respective identity matrices.

The diagonal matrix H_U is due to the onsite repulsion U which counts the double occupancy of a site.

The complete Hamiltonian matrix:

$$H = \mathcal{I}_\downarrow \otimes H_\uparrow \oplus H_\downarrow \otimes \mathcal{I}_\uparrow \oplus H_U, \qquad (6.4)$$

can be obtained as follows

$$\mathcal{I}_\downarrow \otimes H_\uparrow = \begin{pmatrix}
0 & -t & 0 & -t & 0 & 0 & 0 & 0 & 0 & 0 & 0 & 0 & 0 & 0 & 0 & 0 \\
-t & 0 & -t & 0 & 0 & 0 & 0 & 0 & 0 & 0 & 0 & 0 & 0 & 0 & 0 & 0 \\
0 & -t & 0 & -t & 0 & 0 & 0 & 0 & 0 & 0 & 0 & 0 & 0 & 0 & 0 & 0 \\
-t & 0 & -t & 0 & 0 & 0 & 0 & 0 & 0 & 0 & 0 & 0 & 0 & 0 & 0 & 0 \\
0 & 0 & 0 & 0 & 0 & -t & 0 & -t & 0 & 0 & 0 & 0 & 0 & 0 & 0 & 0 \\
0 & 0 & 0 & 0 & -t & 0 & -t & 0 & 0 & 0 & 0 & 0 & 0 & 0 & 0 & 0 \\
0 & 0 & 0 & 0 & 0 & -t & 0 & -t & 0 & 0 & 0 & 0 & 0 & 0 & 0 & 0 \\
0 & 0 & 0 & 0 & -t & 0 & -t & 0 & 0 & 0 & 0 & 0 & 0 & 0 & 0 & 0 \\
0 & 0 & 0 & 0 & 0 & 0 & 0 & 0 & 0 & -t & 0 & -t & 0 & 0 & 0 & 0 \\
0 & 0 & 0 & 0 & 0 & 0 & 0 & 0 & -t & 0 & -t & 0 & 0 & 0 & 0 & 0 \\
0 & 0 & 0 & 0 & 0 & 0 & 0 & 0 & 0 & -t & 0 & -t & 0 & 0 & 0 & 0 \\
0 & 0 & 0 & 0 & 0 & 0 & 0 & 0 & -t & 0 & -t & 0 & 0 & 0 & 0 & 0 \\
0 & 0 & 0 & 0 & 0 & 0 & 0 & 0 & 0 & 0 & 0 & 0 & 0 & -t & 0 & -t \\
0 & 0 & 0 & 0 & 0 & 0 & 0 & 0 & 0 & 0 & 0 & 0 & -t & 0 & -t & 0 \\
0 & 0 & 0 & 0 & 0 & 0 & 0 & 0 & 0 & 0 & 0 & 0 & 0 & -t & 0 & -t \\
0 & 0 & 0 & 0 & 0 & 0 & 0 & 0 & 0 & 0 & 0 & 0 & -t & 0 & -t & 0
\end{pmatrix}$$

Hilbert Space of Model Hamiltonians

$$H_\downarrow \otimes \mathcal{I}_\uparrow = \begin{pmatrix}
0 & 0 & 0 & 0 & -t & 0 & 0 & 0 & 0 & 0 & 0 & 0 & -t & 0 & 0 & 0 \\
0 & 0 & 0 & 0 & 0 & -t & 0 & 0 & 0 & 0 & 0 & 0 & 0 & -t & 0 & 0 \\
0 & 0 & 0 & 0 & 0 & 0 & -t & 0 & 0 & 0 & 0 & 0 & 0 & 0 & -t & 0 \\
0 & 0 & 0 & 0 & 0 & 0 & 0 & -t & 0 & 0 & 0 & 0 & 0 & 0 & 0 & -t \\
-t & 0 & 0 & 0 & 0 & 0 & 0 & 0 & -t & 0 & 0 & 0 & 0 & 0 & 0 & 0 \\
0 & -t & 0 & 0 & 0 & 0 & 0 & 0 & 0 & -t & 0 & 0 & 0 & 0 & 0 & 0 \\
0 & 0 & -t & 0 & 0 & 0 & 0 & 0 & 0 & 0 & -t & 0 & 0 & 0 & 0 & 0 \\
0 & 0 & 0 & -t & 0 & 0 & 0 & 0 & 0 & 0 & 0 & -t & 0 & 0 & 0 & 0 \\
0 & 0 & 0 & 0 & -t & 0 & 0 & 0 & 0 & 0 & 0 & 0 & -t & 0 & 0 & 0 \\
0 & 0 & 0 & 0 & 0 & -t & 0 & 0 & 0 & 0 & 0 & 0 & 0 & -t & 0 & 0 \\
0 & 0 & 0 & 0 & 0 & 0 & -t & 0 & 0 & 0 & 0 & 0 & 0 & 0 & -t & 0 \\
0 & 0 & 0 & 0 & 0 & 0 & 0 & -t & 0 & 0 & 0 & 0 & 0 & 0 & 0 & -t \\
-t & 0 & 0 & 0 & 0 & 0 & 0 & 0 & -t & 0 & 0 & 0 & 0 & 0 & 0 & 0 \\
0 & -t & 0 & 0 & 0 & 0 & 0 & 0 & 0 & -t & 0 & 0 & 0 & 0 & 0 & 0 \\
0 & 0 & -t & 0 & 0 & 0 & 0 & 0 & 0 & 0 & -t & 0 & 0 & 0 & 0 & 0 \\
0 & 0 & 0 & -t & 0 & 0 & 0 & 0 & 0 & 0 & 0 & -t & 0 & 0 & 0 & 0
\end{pmatrix}$$

$$H_U = \begin{pmatrix}
3U & 0 & 0 & 0 & 0 & 0 & 0 & 0 & 0 & 0 & 0 & 0 & 0 & 0 & 0 & 0 \\
0 & 2U & 0 & 0 & 0 & 0 & 0 & 0 & 0 & 0 & 0 & 0 & 0 & 0 & 0 & 0 \\
0 & 0 & 2U & 0 & 0 & 0 & 0 & 0 & 0 & 0 & 0 & 0 & 0 & 0 & 0 & 0 \\
0 & 0 & 0 & 2U & 0 & 0 & 0 & 0 & 0 & 0 & 0 & 0 & 0 & 0 & 0 & 0 \\
0 & 0 & 0 & 0 & 2U & 0 & 0 & 0 & 0 & 0 & 0 & 0 & 0 & 0 & 0 & 0 \\
0 & 0 & 0 & 0 & 0 & 3U & 0 & 0 & 0 & 0 & 0 & 0 & 0 & 0 & 0 & 0 \\
0 & 0 & 0 & 0 & 0 & 0 & 2U & 0 & 0 & 0 & 0 & 0 & 0 & 0 & 0 & 0 \\
0 & 0 & 0 & 0 & 0 & 0 & 0 & 2U & 0 & 0 & 0 & 0 & 0 & 0 & 0 & 0 \\
0 & 0 & 0 & 0 & 0 & 0 & 0 & 0 & 2U & 0 & 0 & 0 & 0 & 0 & 0 & 0 \\
0 & 0 & 0 & 0 & 0 & 0 & 0 & 0 & 0 & 2U & 0 & 0 & 0 & 0 & 0 & 0 \\
0 & 0 & 0 & 0 & 0 & 0 & 0 & 0 & 0 & 0 & 3U & 0 & 0 & 0 & 0 & 0 \\
0 & 0 & 0 & 0 & 0 & 0 & 0 & 0 & 0 & 0 & 0 & 2U & 0 & 0 & 0 & 0 \\
0 & 0 & 0 & 0 & 0 & 0 & 0 & 0 & 0 & 0 & 0 & 0 & 2U & 0 & 0 & 0 \\
0 & 0 & 0 & 0 & 0 & 0 & 0 & 0 & 0 & 0 & 0 & 0 & 0 & 2U & 0 & 0 \\
0 & 0 & 0 & 0 & 0 & 0 & 0 & 0 & 0 & 0 & 0 & 0 & 0 & 0 & 2U & 0 \\
0 & 0 & 0 & 0 & 0 & 0 & 0 & 0 & 0 & 0 & 0 & 0 & 0 & 0 & 0 & 3U
\end{pmatrix}$$

$$H = \begin{pmatrix}
3U & -t & 0 & -t & -t & 0 & 0 & 0 & 0 & 0 & 0 & 0 & -t & 0 & 0 & 0 \\
-t & 2U & -t & 0 & 0 & -t & 0 & 0 & 0 & 0 & 0 & 0 & 0 & -t & 0 & 0 \\
0 & -t & 2U & -t & 0 & 0 & -t & 0 & 0 & 0 & 0 & 0 & 0 & 0 & -t & 0 \\
-t & 0 & -t & 2U & 0 & 0 & 0 & -t & 0 & 0 & 0 & 0 & 0 & 0 & 0 & -t \\
-t & 0 & 0 & 0 & 2U & -t & 0 & -t & -t & 0 & 0 & 0 & 0 & 0 & 0 & 0 \\
0 & -t & 0 & 0 & -t & 3U & -t & 0 & 0 & -t & 0 & 0 & 0 & 0 & 0 & 0 \\
0 & 0 & -t & 0 & 0 & -t & 2U & -t & 0 & 0 & -t & 0 & 0 & 0 & 0 & 0 \\
0 & 0 & 0 & -t & -t & 0 & -t & 2U & 0 & 0 & 0 & -t & 0 & 0 & 0 & 0 \\
0 & 0 & 0 & 0 & -t & 0 & 0 & 0 & 2U & -t & 0 & -t & -t & 0 & 0 & 0 \\
0 & 0 & 0 & 0 & 0 & -t & 0 & 0 & -t & 2U & -t & 0 & 0 & -t & 0 & 0 \\
0 & 0 & 0 & 0 & 0 & 0 & -t & 0 & 0 & -t & 3U & -t & 0 & 0 & -t & 0 \\
0 & 0 & 0 & 0 & 0 & 0 & 0 & -t & -t & 0 & -t & 2U & 0 & 0 & 0 & -t \\
-t & 0 & 0 & 0 & 0 & 0 & 0 & 0 & -t & 0 & 0 & 0 & 2U & -t & 0 & -t \\
0 & -t & 0 & 0 & 0 & 0 & 0 & 0 & 0 & -t & 0 & 0 & -t & 2U & -t & 0 \\
0 & 0 & -t & 0 & 0 & 0 & 0 & 0 & 0 & 0 & -t & 0 & 0 & -t & 2U & -t \\
0 & 0 & 0 & -t & 0 & 0 & 0 & 0 & 0 & 0 & 0 & -t & -t & 0 & -t & 3U
\end{pmatrix}.$$

The Hamiltonian matrix obtained can be diagonalized completely to get all the eigenvalues and eigenvectors using standard algorithms. For that, all the matrix elements need to be stored. But if the matrix is sparse and only a few lower eigenvalues and eigenvectors are required, we can use the iterative methods like Lanczos [18] or Davidson method [19].

The generation of the Hamiltonian matrix in this way could also be done for other similar models like extended Hubbard models, Anderson impurity model, etc.

REFERENCES

[1] Hubbard J. (1963), Electron correlations in narrow energy bands. *Proceedings of the Royal Society of London,* Series A, 276, 238.

[2] Lieb E. H. and Wu F. Y. (1968), Absence of Mott Transition in an Exact Solution of the Short-Range, One-Band Model in One Dimension. *Physical Review Letters*, 20.

[3] Caffarel M. and Krauth W. (1994), Exact diagonalization approach to correlated fermions in infinite dimensions: Mott transition and superconductivity. *Physical Review Letters*, 72, 1545.

[4] Dagotto E. (1994), Correlated electrons in high-temperature superconductors. *Review of Modern Physics*, 66, 763.

[5] Sarma D. D., Ramasesha S., Taraphder A. (1989), Hole pairing within an extended Anderson impurity model applicable to the high-T_c cuprates. *Physical Review B*, 39, 12286.

[6] Callaway J., Chen D. P., Kanhere D. G. and Li Q. (1990), Small-cluster calculations for the simple and extended Hubbard models. *Physical Review B*, 42, 465.

[7] Bednorz J. G. and Müller K. A. (1986), Possible high T_c superconductivity in the Ba-La-Cu-O system. *Zeitschrift für Physik B Condensed Matter*, 64, 189.

[8] Metzner W., Vollhardt D. (1989), Correlated Lattice Fermions in $d = \infty$ Dimensions. *Physical Review Letters*, 62, 324.

[9] Georges A., Kotliar G. (1992), Hubbard model in infinite dimensions. *Physical Review B*, 45, 6479.

[10] Anderson P. W. (1961), Localized Magnetic States in Metals. *Physical Review*, 124, 41.

[11] Hirsch J. E. and Fye R. M. (1986), Monte Carlo Method for Magnetic Impurities in Metals. *Physical Review Letters*, 56, 2521.

[12] Bulla R. (1999), Zero Temperature Metal-Insulator Transition in the Infinite-Dimensional Hubbard Model. *Physical Review Letters*, 83, 136.

[13] Weiße A. and Fehske H. (2008), *In Computational Many-Particle Physics, Lecture Notes in Physics.* Vol. 739, edited by H. Fehske, R. Schneider, and A. Weiße (Springer, Heidelberg, 2008), pp. 529-544.

[14] Jafari S. A. (2008), Introduction to Hubbard Model and Exact Diagonalization. *Iranian Journal of Physics Research*, 8, 113.

[15] Sandvik A. W. (2010), Computational Studies of Quantum Spin Systems. *AIP Conference Proceedings*, 1297, 135.

[16] Lin H. Q. and Gubernatis J. E. (1993), Exact Diagonalization Methods for Quantum Systems. *Computers in Physics*, 7, 400.

[17] Sharma M. and Ahsan M. A. H. (2015), Organization of the Hilbert space for exact diagonalization of Hubbard model. *Computer Physics Communications*, 193, 19-29.

[18] Lanczos C. (1950), An iteration method for the solution of the eigenvalue problem of linear differential and integral operators. *Journal of Research of the National Bureau of Standards*, 45, 255.

[19] Davidson E. R. (1975), The iterative calculation of a few of the lowest eigenvalues and corresponding eigenvectors of large real-symmetric matrices. *Journal of Computational Physics*, 17, 87.

In: Hilbert Spaces: Properties and Applications
Editor: Le Bin Ho

ISBN: 978-1-53616-633-0
© 2020 Nova Science Publishers, Inc.

Chapter 7

ENLARGED HILBERT SPACES AND APPLICATIONS IN QUANTUM PHYSICS

Le Bin Ho[*]
Department of Physics, Kindai University, Higashiosaka City, Japan
Ho Chi Minh City Institute of Physics, VAST,
Ho Chi Minh City, Vietnam

Abstract

In this chapter, we conceptually introduce mathematical foundations and applications of enlarged Hilbert spaces. We first introduce the basic concepts of the enlarged Hilbert spaces for both cases of pure and mixed quantum states. We also discuss how to implement such an enlarged system in various physical platforms. We later focused on applications of the enlarged Hilbert spaces in quantum physics. Topics discussed in this context including noncausal transformation problem, real quantum bits, and two-state vector formalism. Finally, we give a connection of the enlarged Hilbert spaces to the quantum simulation.

Keywords: enlarged Hilbert spaces, Trotter technique, Mølmer-Sørensen gate, noncausal transformation problem, real quantum bits (Rebit), two-state vector formalism (TSVF)

---

[*]Corresponding Author's E-mail: binho@kindai.ac.jp.

1. INTRODUCTION

Physics always deals with mathematics in the sense that any physical observable is mathematically represented by an operator acting on an appropriate state. Concretely, a physical quantity Q, such as position, momentum, energy, and other quantities, is represented by a mathematical operator $\boldsymbol{Q}$, such that can be able to perform basic algebraic operations on it [1]. The spectrum of $\boldsymbol{Q}$ is the possible values of the measurable quantity in Q. In classical mechanics, all $\boldsymbol{Q}$ commute. In quantum mechanics, however, they do not always commute. If the state of a system is represented by a "ket" vector in a Hilbert space, i.e., $|\psi\rangle$, then an expectation value of $\boldsymbol{Q}$ is a linear functional: $\langle\psi|\boldsymbol{Q}|\psi\rangle$, which is also in the Hilbert space. In other words, a scalar product of the state or any operations on it must turn the algebra into the Hilbert space. As a result, we can use this Hilbert space to represent the algebra on it.

Recently, quantum simulators have been proposed and widely used [1–3]. It is a one-to-one mapping between a quantum system (usually named as a *simulated system*) to a given mathematical model (named as *simulator system*), which is more controllable for reproducing the dynamics of the quantum system [1, 2]. The corresponding Hilbert spaces are named as simulated Hilbert space and simulator Hilbert space, respectively. In general, they are different. If the simulator Hilbert space is larger, it is sometimes, named as *enlarged Hilbert space*. Such an *enlarged Hilbert space formalism* for quantum simulator has been proposed by Solano and his colleagues and is extensively studied recently both theoretically and experimentally [4–13]. This chapter introduces the concept of the enlarged Hilbert space in quantum mechanics as well as its applications. To this end, we will present the mapping processes to form enlarged Hilbert spaces for both pure quantum states and mixed quantum states. We also discuss how to implement such the enlarged system in various physics platforms. Later on, we introduce various applications of the enlarged Hilbert space formalism, including noncausal transformation problem, real quantum bits, and two-state vector formalism. Finally, we concuss the chapter with some further aspects.

---

[1] Some basic algebraic operations such as sum, multiply, scale, and so on.

2. ENLARGED HILBERT SPACE IN QUANTUM MECHANICS

In quantum mechanics, a quantum system is usually described by quantum states which live in complex Hilbert spaces. We denote this quantum system is the original system ($\mathcal{OS}$), and its state is the original quantum state (OQS). For simplicity, we also assume that all the OQSs have the same original Hilbert space. Let us assume that $\mathcal{H}_n$ is the Hilbert space of n−dimensional complex vectors, while $\mathcal{L}(\mathcal{H}_n)$ represents the Hilbert space of $n \times n$ matrices. A pure OQS will live in $\mathcal{H}_n$, while a mixed OQS belongs to $\mathcal{L}(\mathcal{H}_n)$.

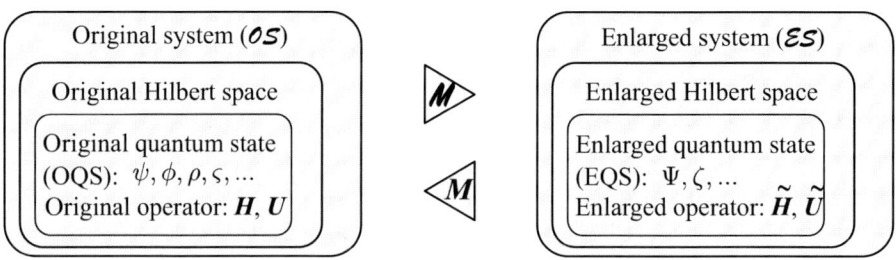

Figure 1. Schematic of the mapping process from an original system onto an enlarged system. The mapping $\mathcal{M}$ will map the original quantum states or operators onto the enlarged ones while the operator M will decoding back to the original system.

We will introduce a mapping process that maps an $\mathcal{OS}$ onto an enlarged system ($\mathcal{ES}$) as depicted in Fig. 1, for both cases of pure OQSs and mixed OQSs as following.

2.1. Mapping Process for Pure OQSs

2.1.1. Enlarged Quantum State

In the $\mathcal{OS}$, let us consider two pure OQSs ψ and ϕ, which live in $\mathcal{H}_n$ Hilbert space. In general, these states are different. We introduce a mapping process $\mathcal{M} : \mathcal{H}_n \to \mathcal{H}_2 \otimes \mathcal{H}_n$ from the original Hilbert space $\mathcal{H}_n$ to an enlarged Hilbert space $\mathcal{H}_2 \otimes \mathcal{H}_n$ that maps both ψ and ϕ onto an enlarged quantum state (EQS)

Ψ, such that [12, 14]

$$\left.\begin{matrix}\psi\\ \phi\end{matrix}\right\} \xrightarrow{\mathcal{M}} \Psi = \frac{1}{2}\begin{pmatrix}\psi+\phi\\ \psi-\phi\end{pmatrix}. \tag{2.1}$$

This mapping is based on the fact that the OQSs can always be expressed as

$$\psi = \frac{1}{2}\Big[(\psi+\phi) + (\psi-\phi)\Big], \text{ and } \phi = \frac{1}{2}\Big[(\psi+\phi) - (\psi-\phi)\Big]. \tag{2.2}$$

As we can see from the mapping, an extra two-dimensional system (extra qubit) has been added to the $\mathcal{OS}$ to form the $\mathcal{ES}$. The EQS can be recast as

$$\Psi = \frac{1}{2}\Big[|0\rangle \otimes (\psi+\phi) + |1\rangle \otimes (\psi-\phi)\Big], \tag{2.3}$$

where $\{|0\rangle, |1\rangle\}$ are two computation bases of the extra qubit.

The EQS in the $\mathcal{ES}$ can be decoded back to the two OQSs in the $\mathcal{OS}$ by

$$\psi = \boldsymbol{M}\Psi, \text{ and} \tag{2.4}$$
$$\phi = \boldsymbol{M}(\sigma_z \otimes \boldsymbol{I}_n)\Psi, \tag{2.5}$$

where $\boldsymbol{M} \equiv \begin{pmatrix}1 & 1\end{pmatrix} \otimes \boldsymbol{I}_n$ is a decoding operator in the enlarged Hilbert space $\mathcal{L}(\mathcal{H}_2 \otimes \mathcal{H}_n)$, σ_i ($i = x, y, z$) is a Pauli matrix, $\boldsymbol{I}_n \in \mathcal{L}(\mathcal{H}_n)$ is the identity matrix [14].

2.1.2. Enlarged Schrödinger Equation

In the $\mathcal{OS}$, the OQS ψ (and also ϕ) satisfies the time-dependent Schrödinger equation:

$$i\hbar\frac{\partial \psi(t)}{\partial t} = \boldsymbol{H}\psi(t), \tag{2.6}$$

where i is the imaginary unit, $\hbar = h/2\pi$ is the reduced Planck constant, and $\boldsymbol{H}$ is the system Hamiltonian in the original Hilbert space $\mathcal{H}_n$. Throughout this chapter, we consider the case that $\boldsymbol{H}$ is time-independent. When the initial condition $\psi(t_i)$ is known, then the OQS at time t can be determined entirely throughout Eq. (2.6).

Similarly, in the $\mathcal{ES}$, the EQS could be governed by an enlarged Schrödinger equation that satisfies [8]

$$i\hbar\frac{\partial \Psi(t)}{\partial t} = \tilde{\boldsymbol{H}}\Psi(t). \tag{2.7}$$

Here $\tilde{H}$ is the enlarged Hamiltonian in the enlarged Hilbert space $\mathcal{H}_2 \otimes \mathcal{H}_n$, which is also time-independent. To find $\tilde{H}$, from Eq. (2.4), we note that

$$\psi(t) = M\Psi(t). \tag{2.8}$$

Therefore, if $\Psi(t)$ is the solution of Eq. (2.7) then $M\Psi(t)$ is the solution of Eq. (2.6). Substituting Eq. (2.8) to Eq. (2.6), we obtain

$$i\hbar \frac{\partial M\Psi(t)}{\partial t} = HM\Psi(t). \tag{2.9}$$

Since M is assumed to be time-independent, Eq. (2.9) is recast as

$$Mi\hbar \frac{\partial \Psi(t)}{\partial t} = HM\Psi(t). \tag{2.10}$$

Now, substituting Eq. (2.7) to Eq. (2.10), we have

$$M\tilde{H}\Psi(t) = HM\Psi(t), \tag{2.11}$$

Alternatively, $M\tilde{H} = HM$. One possible solution for this equation can be chosen as

$$\tilde{H} = \begin{pmatrix} D & B \\ B & D \end{pmatrix} = I_2 \otimes D + \sigma_x \otimes B, \tag{2.12}$$

where $B \in \mathcal{L}(\mathcal{H}_n)$ is an arbitrary $n \times n$ matrix and $D \equiv H - B$, and I_2 is the 2-dimensional identity matrix.

2.1.3. Initial Condition

We now discuss the initial EQS $\Psi(t_i)$. In general, it is given as

$$\Psi(t_i) = \frac{1}{2} \begin{pmatrix} \psi(t_i) + \phi(t_i) \\ \psi(t_i) - \phi(t_i) \end{pmatrix}. \tag{2.13}$$

It implies that the choice of the initial EQS depends on initial conditions of all the OQSs. A proper choice depends on each problem, which we will discuss for individual case later on.

2.2. Mapping Process for Mixed OQSs

2.2.1. Enlarged Quantum State

We consider two mixed OQSs ρ and ς in the $\mathcal{OS}$. Similarly, as Sec. 2.1, these states are generally different; however, have the same Hilbert space $\mathcal{L}(\mathcal{H}_n)$. We introduce a mapping process $\mathcal{M} : \mathcal{L}(\mathcal{H}_n) \to \mathcal{L}(\mathcal{H}_2 \otimes \mathcal{H}_n)$ from the original Hilbert space $\mathcal{L}(\mathcal{H}_n)$ to an enlarged Hilbert space $\mathcal{L}(\mathcal{H}_2 \otimes \mathcal{H}_n)$ that maps both ρ and ς onto an enlarged quantum state ζ in the following [14, 15]

$$\zeta = \frac{1}{2} \begin{pmatrix} \rho & 0_n \\ 0_n & \varsigma \end{pmatrix}, \qquad (2.14)$$

where $0_n \in \mathcal{L}(\mathcal{H}_n)$ is an $n \times n$ zero matrix. We add $1/2$ to normalize the EQS ζ, i.e., $\text{trace}(\zeta) = 1$. In this mapping, the extra qubit system is adding the $\mathcal{OS}$ is such a way that:

$$\zeta = \frac{1}{2}\Big[|0\rangle\langle 0| \otimes \rho + |1\rangle\langle 1| \otimes \varsigma\Big]. \qquad (2.15)$$

The states in the $\mathcal{OS}$ can be decoded by the inversions [15]

$$\rho = 2\boldsymbol{M}\zeta\boldsymbol{N}, \text{ and } \varsigma = 2\boldsymbol{M}\zeta(\sigma_x \otimes \boldsymbol{I}_n)\boldsymbol{N}, \qquad (2.16)$$

where $\boldsymbol{M} = \begin{pmatrix} 1 & 1 \end{pmatrix} \otimes \boldsymbol{I}_n$ and $\boldsymbol{N} = \begin{pmatrix} 1 \\ 0 \end{pmatrix} \otimes \boldsymbol{I}_n$.

2.2.2. Enlarged Master Equation

The evolutions in time of mixed states are governed by the so-called von Neumann equations. In the Schrödinger picture, the von Neumann equations for ρ and ς states that

$$\frac{d\rho(t)}{dt} = -\frac{i}{\hbar}[\boldsymbol{H}, \rho(t)], \text{ and} \qquad (2.17)$$

$$\frac{d\varsigma(t)}{dt} = -\frac{i}{\hbar}[\boldsymbol{H}, \varsigma(t)], \qquad (2.18)$$

where the brackets denote a commutator. Therein, we consider the simplest case where the $\mathcal{OS}$ Hamiltonians that govern ρ and ς are the same.

Enlarged Hilbert Spaces and Applications in Quantum Physics 159

To find the enlarged master equation, we consider the derivation of Eq. (2.14) as

$$\frac{d\zeta(t)}{dt} = \frac{1}{2}\begin{pmatrix} \frac{d\rho(t)}{dt} & 0_n \\ 0_n & \frac{d\varsigma(t)}{dt} \end{pmatrix}. \tag{2.19}$$

Substituting Eqs. (2.17, 2.18) to Eq. (2.19), we have

$$\frac{d\zeta(t)}{dt} = \frac{-i}{2\hbar}\begin{pmatrix} [\boldsymbol{H}, \rho(t)] & 0_n \\ 0_n & [\boldsymbol{H}, \varsigma(t)] \end{pmatrix}$$

$$= \frac{-i}{\hbar}[\tilde{\boldsymbol{H}}, \zeta(t)], \tag{2.20}$$

where we have defined the enlarged Hamiltonian $\tilde{\boldsymbol{H}} \equiv \boldsymbol{I}_2 \otimes \boldsymbol{H}$.

3. IMPLEMENTATION

In this section, we discuss how to implement the $\mathcal{ES}$ in real physical platforms.

3.1. Enlarged Time-Evolution Unitary Operator

The enlarged Schrödinger equation Eq. (2.7) can be recast as

$$\Psi(t) = \tilde{\boldsymbol{U}}(t, t_i)\Psi(t_i), \tag{3.1}$$

where the enlarged time-evolution unitary operator is given as

$$\tilde{\boldsymbol{U}}(t, t_i) = \mathcal{T}\exp\left[-\frac{i}{\hbar}\int_{t_i}^{t}\tilde{\boldsymbol{H}}d\tau\right]$$

$$= \mathcal{T}\exp\left[-\frac{i}{\hbar}(t - t_i)\tilde{\boldsymbol{H}}\right], \tag{3.2}$$

where $\mathcal{T}$ represents the time-ordering operator. Similarly, the enlarged von Neumann equation is recast as

$$\zeta(t) = \tilde{\boldsymbol{U}}(t, t_i)\zeta(t_i)\tilde{\boldsymbol{U}}^\dagger(t, t_i). \tag{3.3}$$

Our concern is to implement such enlarged time-evolution unitary gate $\tilde{\boldsymbol{U}}(t, t_i)$ such that it can be simulated in real physical systems such as superconducting circuits [16], ion-traps [4, 17–20], quantum photonics [10], and among others (See [2]).

3.2. Trotter Technique

In many physical systems, the $\mathcal{OS}$ can be a many-body system. In such a case, its Hamiltonian is a summation of subsystem Hamiltonians. Since the $\mathcal{ES}$ can be formed by adding an extra system, thus we can expand the enlarged Hamiltonian in term of subsystem Hamiltonians as

$$\tilde{H} = \sum_i \tilde{H}_i, \qquad (3.4)$$

where $\tilde{H}_i$ are subsystem Hamiltonians. In general, these $\tilde{H}_i$ are nonlocal and non-commuting operators. We introduce the Trotter technique [21, 22] stating that the evolution operator is decomposed as

$$\tilde{U}(t, t_i) = e^{-\frac{i}{\hbar}(t-t_i)\sum_j \tilde{H}_j} \approx \left(\prod_j e^{-\frac{i}{\hbar}(t-t_i)\tilde{H}_j/k}\right)^k, \qquad (3.5)$$

where k is the number of Trotter steps. The more k is large, the more approximation is exactly. With the Trotter technique, each component $e^{-\frac{i}{\hbar}(t-t_i)\tilde{H}_j/k}$ can be implemented separately.

3.3. Entangling Mølmer-SøRensen Gate

We consider a many-body system which consists of N qubits. An entangling Mølmer-Sørensen gate that acting on the system is given as [23, 24]

$$U_{\mathrm{MS}}(\theta, \phi) = e^{-\frac{i\theta}{4}(\cos\phi S_x + \sin\phi S_y)^2}, \qquad (3.6)$$

where $S_{x,y} = \sum_{l=1}^{N} \sigma_{x,y}^{(l)}$, θ and ϕ are two rotating angles [23, 24]. This such Mølmer-Sørense gate has been proposed to solve such kind of spin interaction [24–26]. Their studies discuss on the Kitaev's toric code Hamiltonian, where the Hamiltonian describes four-body interactions of spins [27, 28] or in general, the Hamiltonian can be expressed in terms of tensor products of Pauli matrices of N qubits.

3.4. Three-Step Implementation

Let us assume that the $\mathcal{OS}$ is an N-qubit system, and then the $\mathcal{ES}$ is described by an $(N + 1)$-qubit system. We consider the case that each $\tilde{H}_j$ is nonlocal and can be decomposed into tensor products of Pauli matrices of $(N+1)$-qubit system. In this case, an exponential term $e^{-\frac{i}{\hbar}(t-t_i)\tilde{\boldsymbol{H}}_j/k}$ can be implemented by using entangling Mølmer-Sørensen gates and local single-qubit gates via three steps: (see Fig. 2)

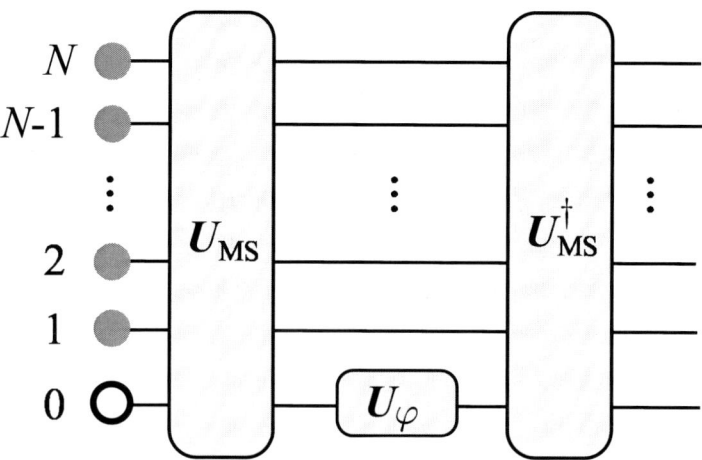

Figure 2. Quantum circuit of the three-step implementation.

Step 1: Operate a Mølmer-Sørensen entangling gate, $\boldsymbol{U}_{\text{MS}}(\theta,\phi)$, to all $(N+1)$ qubits.

Step 2: Apply a local single-qubit gate, $\exp(-i\frac{\varphi}{2}\sigma_\gamma \otimes \boldsymbol{I}_2^{\otimes N})$, to the extra qubit. Here, the phase φ is designed by controlling $2(t-t_i)/k$ and γ is chosen from x, y or z [26].

Step 3: Operate the inverse entangling gate $\boldsymbol{U}_{\text{MS}}^\dagger(\theta,\phi)$ to all $(N+1)$ qubits. This three-step procedure can implement the desired enlarged evolution $e^{-\frac{i}{\hbar}(t-t_i)\tilde{\boldsymbol{H}}_j/k}$ as follows

$$e^{-\frac{i}{\hbar}(t-t_i)\tilde{\boldsymbol{H}}_j/k} = \boldsymbol{U}_{\text{MS}}(\theta,\phi)e^{-i(t-t_i)\sigma_\gamma \otimes \boldsymbol{I}_2^{\otimes N}/k}\boldsymbol{U}_{\text{MS}}^\dagger(\theta,\phi). \quad (3.7)$$

θ and ϕ can be determined when the concreteness form of $\tilde{\boldsymbol{H}}$ is given.

3.5. Example

Let us consider an ensemble of N qubits in the $\mathcal{OS}$ evolves under a magnetic field applies along the y-axis. In the $\mathcal{ES}$, we consider the enlarged evolution which is given by

$$\tilde{U}(t, t_i) = \exp\left[-\frac{i\omega(t - t_i)}{2}\sigma_x \otimes \sum_{l=1}^{N} \sigma_y^{(l)}\right], \quad (3.8)$$

where ω is the magnetic field. The first qubit is the extra system while the rest of N qubits is the $\mathcal{OS}$.

To simulate, we start from two Mølmer-Sørensen gates applied onto all $N + 1$ qubits and one local rotation gate applied on the extra system only:

$$U_{\text{MS}}(\theta, \phi)\exp\left[-\frac{i\omega(t - t_i)}{2}\sigma_z \otimes I_2^{\otimes N}\right]U_{\text{MS}}^{\dagger}(\theta, \phi). \quad (3.9)$$

Using the useful formula $Ue^H U^{\dagger} = e^{UHU^{\dagger}}$ [29], Eq. (3.9) is expanded as

$$\exp\left[-\frac{i\omega(t - t_i)}{2} U_{\text{MS}}(\theta, \phi) (\sigma_z \otimes I_2^{\otimes N}) U_{\text{MS}}^{\dagger}(\theta, \phi)\right]. \quad (3.10)$$

For $\phi = \pi/2$, we have

$$\begin{aligned}U_{\text{MS}}(\theta, 0) &= \exp\left[-\frac{i\theta}{4}\left(\sigma_y \otimes I_2^{\otimes N} + I_2 \otimes \sigma_y \otimes I_2^{\otimes N-1} + \cdots\right)^2\right] \\ &= \exp\left[-\frac{i\theta}{4}\left((N+1)I_2^{\otimes N+1} + 2\sigma_y \otimes \sigma_y \otimes I_2^{\otimes N-1} + \cdots\right)\right].\end{aligned} \quad (3.11)$$

Substituting this into term $U_{\text{MS}}(\theta, \phi)(\sigma_z \otimes I_2^{\otimes N})U_{\text{SM}}^{\dagger}(\theta, \phi)$ in Eq. (3.10), we have

$$U_{\text{MS}}(\theta, \frac{\pi}{2})(\sigma_z \otimes I_2^{\otimes N})U_{\text{MS}}^{\dagger}(\theta, \frac{\pi}{2}) = e^A B e^{-A}, \quad (3.12)$$

where

$$A = (N+1)I_2^{\otimes N+1} + 2\sigma_y \otimes \sigma_y \otimes I_2^{\otimes N-1} + \cdots, \quad (3.13)$$
$$B = \sigma_z \otimes I_2^{\otimes N}. \quad (3.14)$$

Using the Baker-Campbell-Hausdorff relation, we have

$$\begin{aligned} e^{A}Be^{-A} &= B + [A, B] + \frac{1}{2!}[A, [A, B]] + ... \\ &= \cos\theta\, \sigma_x \otimes \sum_l \sigma_y^{(l)} + \sin\theta\, \sigma_z \otimes I^{\otimes N}. \end{aligned} \quad (3.15)$$

We next choose $\theta = 0$, it gives

$$e^{A}Be^{-A} = \sigma_x \otimes \sum_l \sigma_y^{(l)}. \quad (3.16)$$

Substituting Eq. (3.16) to Eq. (3.12), we obtain

$$U_{\mathrm{MS}}(0, \frac{\pi}{2})(\sigma_z \otimes I_2^{\otimes N})U_{\mathrm{MS}}^{\dagger}(0, \frac{\pi}{2}) = \sigma_x \otimes \sum_l \sigma_y^{(l)}. \quad (3.17)$$

Then Eq. (3.10) is recast as

$$\exp\left[-\frac{i\omega(t - t_i)}{2} \sigma_x \otimes \sum_l \sigma_y^{(l)}\right], \quad (3.18)$$

which is the evolution (3.8).

4. APPLICATIONS

Since it was first introduced, several manners of mapping have been suggested for different purposes. For example, a mapping between a quantum state and its spacetime transformation onto an enlarged one could help to solve noncausal transformation problems [12, 13]. Furthermore, a mapping that maps a pair of conjugate wave functions (ψ, ψ^*) onto an enlarged real wave function (real quantum bit) allows us to implement some unphysical operators, such as complex conjugation $\mathcal{K}$, charge conjugation $\mathcal{C}$, and time reversal $\mathcal{T}$ [4–7]. It is also applicable to Majorana particles [4–6], the entanglement monotone [8–11], and direct state measurement [30]. The noncausal problem in the concept of two-state vector formalism also has been considered in an enlarged form [14, 15]. Following we review these applications in detailed.

4.1. Noncausal Transformation Problem

4.1.1. Spacetime Transformation

Let us consider a linear spacetime transformation $(x,t) \to (x',t')$, such that $x' = f(x,t)$ and $t' = g(x,t)$, where f and g are two linear transformation functions. The transformation can be represented in a transfer matrix T as

$$\begin{pmatrix} x' \\ t' \end{pmatrix} = \begin{pmatrix} t_{00} & t_{01} \\ t_{10} & t_{11} \end{pmatrix} \begin{pmatrix} x \\ t \end{pmatrix}, \tag{4.1}$$

where t_{ij} are the components of matrix T. It is a spatial transformation when $T = \text{diag}[1, t_{11}]$, while a time transformation corresponds to $T = \text{diag}[t_{00}, 1]$.

In an $\mathcal{OS}$ that described by a quantum state ψ, the transformation $\psi(x,t) \to \psi(x',t')$ is noncausal and, in general, violates the special relativity [12]. For example, the time transformation $\psi(x,t) \to \psi(x,-t)$ is noncausal, while the spatial transformation $\psi(x,t) \to \psi(-x,t)$ violates the special relativity.

4.1.2. Enlarged System

To solve this problem, Alvarez-Rodriguez et al. [12] introduced a mapping process onto an $\mathcal{ES}$ in such a way that

$$\Psi = \frac{1}{2} \begin{pmatrix} \psi(x,t) + \psi(x',t') \\ \psi(x,t) - \psi(x',t') \end{pmatrix}, \tag{4.2}$$

where we have used $\psi \equiv \psi(x,t)$ and $\phi \equiv \psi(x',t')$ in Eq. (2.1). In this form, the transformation state $\psi(x',t')$ can be achieved in the $\mathcal{ES}$ by applying a physical action such that

$$\psi(x',t') = M(\sigma_z \otimes I_n)\Psi. \tag{4.3}$$

Here $M = \begin{pmatrix} 1 & 1 \end{pmatrix} \otimes I_n$, where n the dimension of the $\mathcal{OS}$. In other words, such a noncausal transformation problem can be solved in the $\mathcal{ES}$.

The dynamic of $\Psi(x,t)$ in the $\mathcal{ES}$ is the same as Eq. (2.7). The initial state $\Psi(x, t_i = 0)$ can be determined entirely by the initial condition $\psi(x,0)$ in the $\mathcal{OS}$ [12].

4.2. Switchable

From the previous mapping method, it can be switched between the dynamic of $\psi(x,t)$ and $\psi(x',t')$ by operating on the enlarged state, such as $\psi(x,t) = M\Psi(x,t)$ and $\psi(x',t') = M(\sigma_z \otimes I)\Psi(x,t)$. As a result, for example, the expectation values $\langle\psi|Q|\psi\rangle$, $\langle\phi|Q|\phi\rangle$, and the correlation $\langle\phi|Q|\psi\rangle$ can be obtained in the $\mathcal{ES}$. Here $\langle x|\psi\rangle = \psi(x,t)$ and $\langle x'|\phi\rangle = \psi(x',t')$.

Recently, Cheng et al. [31] have introduced a mapping between boson and fermion onto an enlarged one that can allow switching between them. The enlarged state is given as

$$\Psi = \frac{1}{\sqrt{2}} \begin{pmatrix} \psi_b \\ \psi_f \end{pmatrix} \tag{4.4}$$

where ψ_b and ψ_f are bosonic and fermionic wave functions, respectively. Therefore, the dynamic of the enlarged system can be switched between the Bose-Einstein and Fermi-Dirac statistics [31].

4.3. Real Quantum Bit (Rebit)

It is well known that a universal quantum computing can be efficiently transformed to circuits that use only the real amplitudes [32, 33], e.g., the Quantum Turing Machine (QTM) model [34]. The state in such transformation is a superposition of rebits, the *real versions* of qubits. Rebits are, recently, widely explored and used in simulating quantum systems [35, 36]. Following, we introduce the concept of the rebit in the $\mathcal{ES}$ and its applications.

4.3.1. Rebits in the Enlarged System

Given a complex quantum state in the $\mathcal{OS}$ such that

$$\psi = \psi^r + i\psi^i \tag{4.5}$$

where $\psi^r \equiv \text{Real}[\psi]$ and $\psi^i \equiv \text{Im}[\psi]$ are real and imaginary parts of the state ψ, respectively. Rebit can be defined through a mapping

$$\Psi = \begin{pmatrix} \psi^r \\ \psi^i \end{pmatrix} = \begin{pmatrix} \psi^r_1 \\ \vdots \\ \psi^r_n \\ \psi^i_1 \\ \vdots \\ \psi^i_n \end{pmatrix}, \qquad (4.6)$$

where in Eq. (2.1) we have chosen $\psi \equiv 2\psi^r$ and $\phi \equiv 2\psi^i$. In this case, the decoding operator $M = \begin{pmatrix} 1 & i \end{pmatrix} \otimes I_n$.

The evolution of the enlarged quantum state Ψ in the $\mathcal{ES}$ can be described by Eq. (2.7), where $\tilde{H}$ can be chosen as $\tilde{H} = I_2 \otimes \text{Re}[H] - i\sigma_y \otimes \text{Im}[H]$ [4,8]. We emphasize that any operator Q in the $\mathcal{OS}$ can be mapped onto $\tilde{Q}$ in the $\mathcal{ES}$ in the same way, i.e., $\tilde{Q} = I_2 \otimes \text{Re}[Q] - i\sigma_y \otimes \text{Im}[Q]$ [4]. For more references see [37].

4.3.2. $\mathcal{K}, \mathcal{C}, \mathcal{T}$ Unphysical Operators

In quantum mechanics, there are some unphysical operators which cannot be implemented in physical systems [38]. Here we show that by using rebits, these unphysical operators, including complex conjugation $\mathcal{K}$, charge conjugation $\mathcal{C}$, and time reversal $\mathcal{T}$, can be implemented in the $\mathcal{ES}$.

We first consider the complex conjugation $\mathcal{K}$, such that

$$\psi^* = \mathcal{K}\psi = \psi^r - i\psi^i. \qquad (4.7)$$

$\mathcal{K}$ is an antilinear operator or not a physical operator because a linear operator is unable to map a wave function into its complex conjugate. Thus, it cannot be implemented (by any physical system) in the $\mathcal{OS}$. However, it can be done in the $\mathcal{ES}$, such that

$$\begin{aligned} \psi^* &= M(\sigma_z \otimes I_n)\Psi \\ &= \begin{pmatrix} 1 & i \end{pmatrix} \otimes I_n (\sigma_z \otimes I_n) \begin{pmatrix} \psi^r \\ \psi^i \end{pmatrix} \\ &= \psi^r - i\psi^i. \end{aligned} \qquad (4.8)$$

The operator $(\sigma_z \otimes I_n)$ can be realized by operating σ_z into the extra system while leaving the $\mathcal{OS}$ alone. Furthermore, M can be done in a reverse mapping process (see Ref. [4].)

Next, we consider the charge conjugation $\mathcal{C}$, which exchanges particles and antiparticles. Here we only focus on the Dirac field. The operation of the charge conjugation into the quantum state is given as

$$\psi_c = \mathcal{C}\psi = \mathscr{C}\Gamma^0 \psi^*, \tag{4.9}$$

where $\mathscr{C}$ is a unitary matrix satisfying $\mathscr{C}(\Gamma^\mu)^T \mathscr{C}^{-1} = -\Gamma^\mu$, where $\Gamma^\mu, \mu = \{0, 1, \cdots, n-1\}$ are Dirac matrices in n-dimension [39]. The enlarged form of $\mathscr{C}\Gamma^0$ is given as

$$\widetilde{\mathscr{C}\Gamma^0} = I_2 \otimes \text{Re}[\mathscr{C}\Gamma^0] - i\sigma_y \otimes \text{Im}[\mathscr{C}\Gamma^0]. \tag{4.10}$$

Substituting Eqs. (4.8, 4.10) into Eq. (4.9), we have

$$\psi_c = M\widetilde{\mathscr{C}\Gamma^0}(\sigma_z \otimes I_n)\Psi. \tag{4.11}$$

As an example, let us focus on the case $n = 2$. In the Dirac representation, $\mathscr{C} = i\sigma_y$ and $\Gamma^0 = \sigma_z$. Then $\mathscr{C}\Gamma^0 = i\sigma_y\sigma_z = -\sigma_x$. In the enlarged form, operator $\widetilde{\mathscr{C}\Gamma^0}$ is given as $-I_2 \otimes \text{Re}[\sigma_x] + i\sigma_y \otimes \text{Im}[\sigma_x] = -I_2 \otimes \sigma_x$. Substituting this enlarged form into Eq. (4.11), we obtain

$$\psi_c = -M(I_2 \otimes \sigma_x)(\sigma_z \otimes I_2)\Psi$$
$$= -M(\sigma_z \otimes \sigma_x)\Psi. \tag{4.12}$$

Again, the operation $(\sigma_z \otimes \sigma_x)$ in the $\mathcal{ES}$ can be implemented by applying σ_z into the extra system and σ_x into the $\mathcal{OS}$. As a result, the charge conjugate can be realized in the $\mathcal{ES}$.

Finally, we consider the time reversal $\mathcal{T}$. The action of $\mathcal{T}$ into a quantum state is given as

$$\psi_\mathcal{T}(t) = \mathcal{T}\psi(t) = \psi^*(-t). \tag{4.13}$$

Again, $\mathcal{T}$ is an antilinear operator, which can not be implemented in the $\mathcal{OS}$. However, in the $\mathcal{ES}$, it can be performed, such that for 2 dimensions [4]

$$\psi_\mathcal{T}(t) = M(\sigma_z \otimes \sigma_z)\Psi. \tag{4.14}$$

These unphysical operators have been experimentally implemented recently [7, 13].

4.3.3. Entanglement Monotone

The implementation some unphysical operators in the $\mathcal{ES}$ using rebits is essentially important to "simulate" or measure some physical phenomena, including Majorana equation [4, 6, 7], Dirac equation [40], and entanglement monotone [8–11]. In this section, we will focus on the implementation of entanglement monotone in the $\mathcal{ES}$.

Entanglement monotone for a pure state ψ is defined as [41, 42]

$$E(\psi) = \mathcal{E}[C(\psi)], \tag{4.15}$$

where the concurrence $C(\psi)$ is given by

$$C(\psi) = \langle \psi | \boldsymbol{Q} | \psi^* \rangle, \tag{4.16}$$

where $\boldsymbol{Q}$ is a Hermitian transformation, $\psi^* = \mathcal{K}\psi$ is the complex conjugate of ψ, $\mathcal{K}$ is the complex conjugate operator. Substituting ψ^* in Eq. (4.8) into Eq. (4.16), the concurrence can be recast in the $\mathcal{ES}$ as [8]

$$C(\psi) = \langle \Psi | \boldsymbol{M}^\dagger \boldsymbol{Q} \boldsymbol{M} (\sigma_z \otimes \boldsymbol{I}_n) | \Psi \rangle. \tag{4.17}$$

Furthermore, we have

$$\boldsymbol{M}^\dagger \boldsymbol{Q} \boldsymbol{M} (\sigma_z \otimes \boldsymbol{I}_n) = (\sigma_z - i\sigma_y) \otimes \boldsymbol{Q}, \tag{4.18}$$

which can be performed in the $\mathcal{ES}$ as a linear combination of Hermitian operators $\sigma_z \otimes \boldsymbol{Q}$ and $\sigma_y \otimes \boldsymbol{Q}$ [8]. This entanglement monotone has been experimentally measured so far in various physics platforms, including photonics systems [10, 11], atomic systems [9], NMR systems [43].

4.3.4. Quantum State Estimation

In quantum mechanics, a complex quantum state plays a crucial role to understand natural phenomena at the quantum scale. The determination of the quantum state thus is crucially demanded and is one of the main tasks in quantum mechanics. There are various methods have been proposed to estimate unknown quantum states, including quantum state tomography (QST) [44–46] and direct quantum state measurement (DSM) [47, 48]. Recently, it is given that a complex quantum state can be estimated better by using rebits [30]. Assume that a quantum state in the $\mathcal{OS}$ can be mapped onto the $\mathcal{ES}$. The enlarged state is a rebit and given as [30]

$$|\Psi\rangle = \begin{pmatrix} \psi_1^r \\ \psi_1^i \\ \psi_2^r \\ \psi_2^i \\ \vdots \end{pmatrix}. \qquad (4.19)$$

Of course, such rebit state can be prepared initially in the $\mathcal{ES}$ [8] and can be experimentally implemented [10, 11].

To estimate this state, the $\mathcal{ES}$ is coupled with a pointer qubit prepared in state $|0\rangle$. After the interaction, the $\mathcal{OS}$ is postselected onto a conjugate basis, where the remain extra qubit and pointer qubit can be measured to give the estimation quantum state. This method can assist in enhancing the accuracy of the precision since the real and imaginary parts of the quantum state can be estimated in one single measurement [30].

4.4. Two-State Vector Formalism

Two-state vector formalism (TSVF) was first introduced by Watanabe [49] and later on rediscovered by Aharonov et al. [50]. In this formalism, the $\mathcal{OS}$ is described by past and future conditions, which is referred to as preselected and postselected states [51, 52]. These two states, together, can affect the statistical results and provide more information about the $\mathcal{OS}$ [53, 54]. In the $\mathcal{OS}$, assume that a preselected state $\psi(t_i)$ is prepared at the initial time (t_i) and a postselected state is given by $\phi(t_f)$ at the final time t_f. In general, the postselected state is different from the evolution of the initial state, $\phi(t_f) \neq \psi(t_f)$. At time t in $[t_i, t_f]$, we define the forward-evolving state $\psi(t)$ which evolves forward in time from t_i to t and the backward-evolving state $\phi(t)$ which evolves backward in time from t_f to t, respectively. Together, they describe the $\mathcal{OS}$ completely.

4.4.1. An Enlarged Two-State

Reznik and Aharonov [55] have defined a type of enlarged state, where they named as "two-state," such that

$$\Psi(t) \equiv |\psi(t)\rangle\langle\phi(t)|, \qquad (4.20)$$

where Ψ_t is the density state. If the Hilbert space of the $\mathcal{OS}$ is $\mathcal{H}_n$, then the two-state Ψ_t lives in the $\mathcal{L}(\mathcal{H}_n)$ Hilbert space. A normalized two-state is given by

$$\Psi(t) = \frac{|\psi(t)\rangle\langle\phi(t)|}{\langle\phi(t)|\psi(t)\rangle}, \quad (4.21)$$

where we always assume that $\langle\phi(t)|\psi(t)\rangle \neq 0$, then $\text{Tr}[\Psi(t)] = 1$. In this type of mapping, the forward- and backward-evolving states in the $\mathcal{OS}$ can be obtained by the decoding [55]

$$\frac{\Psi(t)\Psi^\dagger(t)}{\text{Tr}[\Psi(t)\Psi^\dagger(t)]} = |\psi(t)\rangle\langle\psi(t)|, \text{ and} \quad (4.22)$$

$$\frac{\Psi^\dagger(t)\Psi(t)}{\text{Tr}[\Psi(t)\Psi^\dagger(t)]} = |\phi(t)\rangle\langle\phi(t)|. \quad (4.23)$$

This two-state later is extensively used [56–58].

Recently, Vaidman et al. [59] also defined a so-called "genuine mixed two-state vector"

$$\zeta(t) \equiv \begin{pmatrix} \varsigma(t) & \rho(t) \end{pmatrix}, \quad (4.24)$$

where $\rho(t)$ and $\varsigma(t)$ are density matrices of the forward- and backward-evolving states, respectively. This mapping is $\mathcal{M}: \mathcal{L}(\mathcal{H}_n) \to \mathcal{H}_2 \otimes \mathcal{L}(\mathcal{H}_n)$. In this case, an extra qubit is added:

$$\zeta(t) = (0\ 1) \otimes \rho(t) + (1\ 0) \otimes \varsigma(t). \quad (4.25)$$

It also can be decoded back into the $\mathcal{OS}$ as following

$$\rho(t) = \zeta(t)\begin{pmatrix} 0_n \\ I_n \end{pmatrix}, \text{ and} \quad (4.26)$$

$$\varsigma(t) = \zeta(t)\begin{pmatrix} I_n \\ 0_n \end{pmatrix}. \quad (4.27)$$

4.4.2. Enlarged Weak Values

A measurement at time t which condition by both the preselected and postselected states can be described as follows: (i) at the initial time t_i, we prepare a

Enlarged Hilbert Spaces and Applications in Quantum Physics 171

quantum state as $\psi(t_i)$, (ii) a weak measurement is performed at time t, (iii) the system continues to evolve to the final time t_f and be postselected onto a final state $\phi(t_f)$. The result of such measurement is known as "weak value," which is given by both pre- and postselected state at time t

$$\langle Q(t) \rangle_w = \frac{\langle \phi(t) | Q | \psi(t) \rangle}{\langle \phi(t) | \psi(t) \rangle}, \qquad (4.28)$$

where Q is the measured observable, w stands for "weak" [53].

This measurement, however, is noncausal in time, in the scene that the $\mathcal{OS}$ does not evolve causally from the past to the future. Weak values thus, cannot be monitored continuously in the time since the results of measurements at time t are depended on the final state at time t_f. To solve this problem, we introduce a mapping process that maps both the forward- and backward-evolving states onto an enlarged state such that [14]

$$\Psi(t) = \frac{1}{2} \begin{pmatrix} \psi(t) + \phi(t) \\ \psi(t) - \phi(t) \end{pmatrix}. \qquad (4.29)$$

The $\mathcal{ES}$, therefore, evolves one-way forward in time. This mapping is the same as Eq. (2.1).

The weak value can be recast in the $\mathcal{ES}$ as

$$\begin{aligned}\langle Q(t) \rangle_w &= \frac{\langle \Psi(t) | (\sigma_z \otimes I_n) M^\dagger \, Q \, M | \Psi(t) \rangle}{\langle \Psi(t) | (\sigma_z \otimes I_n) M^\dagger M | \Psi(t) \rangle} \\ &= \frac{\langle \Psi(t) | (\sigma_z + i\sigma_y) \otimes Q | \Psi(t) \rangle}{\langle \Psi(t) | (\sigma_z + i\sigma_y) \otimes I_n | \Psi(t) \rangle}, \end{aligned} \qquad (4.30)$$

which is the expectation value and can be measured continuously in time [14]. Recently, the enlarged weak values for generally mixed pre- and postselected density states also have been proposed [15]. It states that weak values can be measured continuously.

4.5. Enlarged System as a Quantum Simulator

The enlarged system ($\mathcal{ES}$) in the enlarged Hilbert space can be served as a quantum simulator, which is a one-to-one mapping between an original quantum system (the simulated system, $\mathcal{OS}$) to a given mathematical model (the simulator system,) which is more controllable for reproducing the dynamics of the simulated system [1,2], see figure 3.

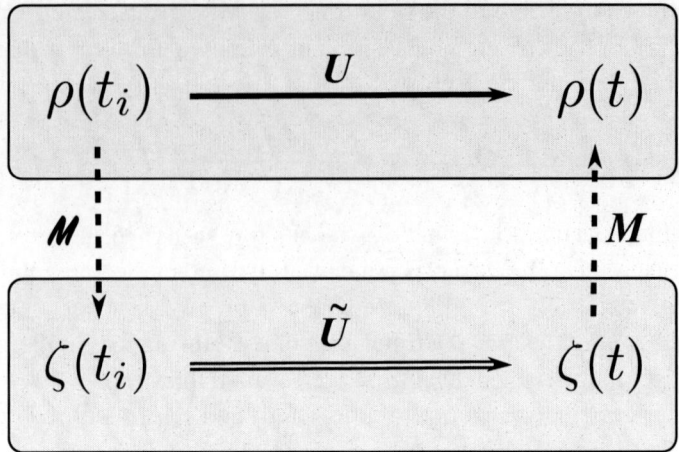

Figure 3. Schematic illustration of the simulated system and the simulator system. A (mixed) simulated system state $\rho(t_i)$ evolves to $\rho(t)$ via the unitary operator U. This process may not be controllable or cannot be accessed in some cases. In such cases, a simulator system could be alternatively used. A mapping $\mathcal{M}$ will map the simulated system onto the simulator system which is able to control: The corresponding simulator state $\zeta(t_i)$ can be prepared, the unitary evolution $\tilde{U}$ can be implemented, and the evolving state $\zeta(t)$ can be experimentally measured. The measurement results in the simulator system provide the dynamic of the simulated system via the decoding process by the operator M.

Typically, the quantum simulator encompasses analog quantum simulation and digital quantum simulation [3]. The main task of quantum simulators is to solve the dynamical time-dependence Schrödinger equation by the fundamental law of nature. Many physical models have already been proposed and demonstrated for quantum simulations (see [2] and Refs. therein).

Notably, in our case, the simulator system is the $\mathcal{ES}$ where it simulates some uncontrollable operators of the $\mathcal{OS}$. As a consequence, by probing of the $\mathcal{ES}$, we also can control and gain information in the $\mathcal{OS}$.

ACKNOWLEDGMENTS

This work was supported by JSPS KAKENHI Grant Number JP19K14620.

CONCLUSION

We have reviewed a concept of enlarged Hilbert spaces and numerous applications to quantum physics. Therein, quantum states in an original system can combine together following a mapping rule to form an enlarged state in an enlarged system. This enlarged system evolves in a different way to the original system, but still, ensures the physical properties of the original system. The enlarged system can be implemented and controlled in various physical platforms that make it valuable to use. As described in Section 4, various physical problems in quantum mechanics have been resolved in the enlarged system, including experimental verification. This work should be useful for guiding further experiments, especially in the scene that there are more advanced technologies today. We also expect that the weak values in the enlarged systems can be experimentally verified in the near future. We believe that the concept of enlarged Hilbert spaces plays an important role to fulfill the understanding of quantum mechanics.

REFERENCES

[1] Feynman R. P. (1982), Simulating Physics with Computers. *International Journal of Theoretical Physic*, 21, 467-488.

[2] Georgescu I. M., Ashhab S. and Nori F. (2014), Quantum simulation. *Reviews of Modern Physics*, 86, 153-185.

[3] Buluta I. and Nori F. (2009), Quantum Simulators. *Science*, 326, 108-111.

[4] Casanova J., Sabin C., Leon C. , Egusquiza I. L., Gerritsma R., Roos C. F., Garcia-Ripoll J. J. and Solano E. (2011), Quantum Simulation of the Majorana Equation and Unphysical Operations. *Physical Review X*, 1, 021018.

[5] Noh C., Rodriguez-Lara B. M. and Angelakis D. G. (2013), Proposal for realization of the Majorana equation in a tabletop experiment. *Physical Review A*, 87, 040102(R).

[6] Rodriguez-Lara B. M. and Moya-Cessa H. M. (2014), Optical simulation of Majorana physics. *Physical Review A*, 89, 015803.

[7] Zhang X., Shen Y., Zhang J., Casanova J., Lamata L., Solano E., Yung M. H., Zhang J. N. and Kim K. (2015), Time reversal and charge conjugation in an embedding quantum simulator. *Nature Communication*, 6, 7917.

[8] DiCandia R., Mejia B., Castillo H., Pedernales J. S., Casanova J. and Solano E. (2013), Embedding Quantum Simulators for Quantum Computation of Entanglement. *Physical Review Letter*, 111, 240502.

[9] Pedernales J. S., DiCandia R., Schindler P., Monz T., Hennrich M., Casanova J. and Solano E. (2014), Entanglement measures in ion-trap quantum simulators without full tomography. *Physical Review A*, 90, 012327.

[10] Loredo J. C., Almeida M. P., Candia R. D., Pedernales J. S., Casanova J., Solan, E. and White A. G. (2016), Measuring Entanglement in a Photonic Embedding Quantum Simulator. *Physical Review Letter*, 116, 070503.

[11] Chen M. C., Wu D., Su Z. E., Cai X. D., Wang X. L., Yang T., Li L., Liu N. L., Lu C. Y. and Pan J. W. (2016), Efficient Measurement of Multiparticle Entanglement with Embedding Quantum Simulator. *Physical Review Letter*, 116, 070502.

[12] Alvarez-Rodriguez U., Casanova J., Lamata L. and Solano E. (2013), Quantum Simulation of Noncausal Kinematic Transformations. *Physical Review Letter*, 111, 090503.

[13] Cheng X. H., Alvarez-Rodriguez U., Lamata L., Chen, X. and Solano E. (2015), Time and spatial parity operations with trapped ions. *Physical Review A*, 92, 022344.

[14] Ho L. B. and Imoto N. (2018), Quantum weak and modular values in enlarged Hilbert spaces. *Physical Review A*, 97, 012112.

[15] Ho L. B. (2019), Continuous-monitoring measured signals bounded by past and future conditions in enlarged quantum systems. *Quantum Information Processing*, 18, 206.

[16] Devoret M. H. and Schoelkopf R. J. (2013), Superconducting Circuits for Quantum Information: An Outlook. *Science*, 339, 1169-1174.

[17] Lanyon B. P., Hempel C., Nigg D., Muller M., Gerritsma R., Zahringer F., Schindler P., Barreiro J. P., Rambach M., Kirchmair G., Hennrich M., Zoller P., Blatt R. and Roos C. F. (2011), Universal Digital Quantum Simulation with Trapped Ions. *Science*, 334, 57-61.

[18] Leibfried D., Blatt R., Monroe C. and Wineland D. (2003), Quantum dynamics of single trapped ions. *Reviews of Modern Physics*, 75, 281.

[19] Batt R. and Roos C. F. (2012), Quantum simulations with trapped ions. *Nature Physics*, 8, 277-284.

[20] Lamata L., Mezzacapo A., Casanova J. and Solano E. (2014), Efficient quantum simulation of fermionic and bosonic models in trapped ions. *EPJ Quantum Technology*, 1, 9.

[21] Trotter H. F. (1959), On the product of semi-groups of operators. *Proceeding of the American Mathematical Society*, 10, 545-551.

[22] Lloyd S. (1996), Universal Quantum Simulators. *Science*, 273, 1073-1078.

[23] Mølmer K. and Sørensen A. (1999), Multiparticle Entanglement of Hot Trapped Ions. *Physical Review Letter*, 82, 1835.

[24] Müller M., Hammerer K., Zhou Y. L., Roos C. F. and Zoller P. (2011), Simulating open quantum systems: from many-body interactions to stabilizer pumping. *New Journal Physics*, 13, 085007.

[25] Barreior J. T., Müller M., Schindler P., Nigg D., Monz T., Chwalla M., Hennrich M., Roos C. F., Zoller P. and Blatt R. (2011), An open-system quantum simulator with trapped ions. *Nature*, 470, 486-491.

[26] Casanova J., Mezzacapo A., Lamata L. and Solano E. (2012), Quantum Simulation of Interacting Fermion Lattice Models in Trapped Ions. *Physical Review Letter*, 108, 190502.

[27] Kitaev A. Y., Shen A. and Vyalyi M. N. (2002), *Classical and quantum computation*. American Mathematical Society Providence.

[28] Kitaev A. Y. (2003), Fault-tolerant quantum computation by anyons. *Annals of Physics*, 303, 2-30.

[29] Dür W., Bremner M. J. and Briegel H. J. (2008), Quantum simulation of interacting high-dimensional systems: The influence of noise. *Physical Review A*, 78, 052325.

[30] Ho L. B. (2019), Improving direct state measurements by using rebits in real enlarged Hilbert spaces. *Physics Letters A*, 383, 289-294.

[31] Cheng X. H., Arrazola I., Pedernales J. S., Lamata L., Chen X. and Solano E. (2017), Switchable particle statistics with an embedding quantum simulator. *Physical Review A*, 95, 022305.

[32] Rudolph T. and Grover L. (2002), arXiv:quant-ph/0210187.

[33] Bennett C. H., Bernstein E., Brassard G. and Vazirani U. (1997), Strengths and Weaknesses of Quantum Computing. *Siam Journal on Computing*, 26, 1510-1523.

[34] Adleman L., Demarrais J. and Huang M-D.A. (1997), Quantum Computability. *Siam Journal on Computing*, 26, 1524-1540.

[35] Delfosse N., Guerin P. A., Bian J. and Raussendorf R. (2015), Wigner Function Negativity and Contextuality in Quantum Computation on Rebits. *Physical Review X*, 5, 021003.

[36] McKague M., Mosca M. and Gisin N. (2009), Simulating Quantum Systems Using Real Hilbert Spaces. *Physical Review Letter*, 102, 020505.

[37] Koh D. E., Niu M. Y. and Yoder T. J. (2018), Quantum simulation from the bottom up: the case of rebits. *Journal of Physics A: Mathematical and Theoretical*, 51, 195302.

[38] Streater R. F. and Wightman A. S. (2000), *PCT, Spin and Statistics and All That*. Princeton University Press, Princeton.

[39] Thaller B. (1992), *The Dirac Equation*. Springer-Verlag, Berlin.

[40] Lee T. E., Alvarez-Rodriguez U., Cheng X. H., Lamata L. and Solano E. (2015), Tachyon physics with trapped ions. *Physical Review A*, 92, 032129.

[41] Hill S. and Wootters W. K. (1997), Entanglement of a Pair of Quantum Bits. *Physical Review Letter*, 78, 5022.

[42] Wootters W. K. (2008), Entanglement of Formation of an Arbitrary State of Two Qubits. *Physical Review Letter*, 80, 2245.

[43] Xin T., Pedernales J. S., Solano E. and Long G. L. (2018), Entanglement measures in embedding quantum simulators with nuclear spins. *Physical Review A*, 97, 022322.

[44] James D. F. V., Kwiat P. G., Munro W. J. and White A. G. (2001), Measurement of qubits. *Physical Review A*, 64, 052312.

[45] Wootters W. K., Fields B. D. (1989), Optimal state-determination by mutually unbiased measurements. *Annals of Physics*, 191, 363-381.

[46] Adamson R. B. A. and Steinberg A. M. (2010), Improving Quantum State Estimation with Mutually Unbiased Bases. *Physical Review Letter*, 105, 030406.

[47] Lundeen J. S., Sutherland B., Patel A., Stewart C., Bamber C. (2011), Direct measurement of the quantum wavefunction. *Nature (London)*, 474, 188-191.

[48] Lundeen J. S. and Bamber C. (2012), Procedure for Direct Measurement of General Quantum States Using Weak Measurement. *Physical Review Letter*, 108, 070402.

[49] Watanabe S. (1955), Symmetry of physical laws. Part III. Prediction and retrodiction. *Reviews of Modern Physics*, 27, 179-186.

[50] Aharonov Y., Bergmann P. G. and Lebowitz J. L. (1964), Time Symmetry in the Quantum Process of Measurement. *Physical Review*, 134, B410-B1416.

[51] *Potentiality, Entanglement and Passion-at-a-Distance*, Quantum Mechanical Studies for A. M. Shimony, edited by Cohen R. S., Horne M. and Stachel J. J. (1997), Kluwer Academic, The Netherlands.

[52] *Time in Quantum Mechanics*, 2nd ed., edited by Muga, J. G., Mayato R. S. and Egusquiza I. (2007), Springer, New York.

[53] Aharonov Y., Albert D. Z. and Vaidman L. (1988), How the result of a measurement of a component of the spin of a spin-1/2 particle can turn out to be 100. *Physical Review Letter*, 60, 1351.

[54] Dressel J., Malik M., Miatto F. M., Jordan A. N. and Boyd R. W. (2014), Colloquium: Understanding quantum weak values: Basics and applications. *Reviews of Modern Physics*, 86, 307.

[55] Reznik B. and Aharonov Y. (1995), Time-symmetric formulation of quantum mechanics. *Physical Review A*, 52, 2538.

[56] Shikano Y. and Hosoya A. (2010), Weak values with decoherence. *Journal of Physics A: Mathematical and Theoretical*, 43, 025304.

[57] Silva R., Guryanova Y., Short A. J., Skrzypczyk P., Brunner N. and Popescu S. (2017), Connecting processes with indefinite causal order and multi-time quantum states. *New Journal Physics*, 19, 103022.

[58] Oreshkov O. and Giarmatzi C. (2016), Causal and causally separable processes. *New Journal Physics*, 18, 093020.

[59] Vaidman L., Ben-Israel A., Dziewior J., Knips L., Weißl M., Meinecke J., Schwemmer C., Ber R. and Weinfurter H. (2017), Weak value beyond conditional expectation value of the pointer readings. *Physical Review A*, 96, 032114.

ABOUT THE EDITOR

Le Bin Ho
Researcher
Department of Physics, Kindai University,
Higashiosaka City, Osaka, Japan
E-mail: binho@kindai.ac.jp; bin262@gmail.com

Le Bin Ho is a junior researcher on the foundation of quantum physics at Kindai University, Japan. He conducted research specifically on the mathematical description and physical interpretation of quantum mechanics, quantum metrology and their applications on quantum computation and quantum information theory. He is serving as a reviewer for the Mathematical Reviews, American Mathematical Society. He also enjoys traveling, sightseeing, and photographs.

INDEX

A

amplitude, 125, 144
analytic solutions, viii, 124, 125, 140
annihilation, 124, 144
arithmetic, 30
atoms, 125

B

Banach spaces, viii, 42, 47, 48, 58, 69, 120
boson(s), 123, 124, 130, 135, 165
boundary value problem, 119
bounded linear operators, 20, 47

C

C*-algebra, 61, 62, 63, 111, 112, 121
calculus, 119, 122
Cauchy problem, 46, 58
classes, viii, 20, 94, 97, 98
classical mechanics, 154
Clifford algebra, 119
closed ball, 44, 52
closure, 7, 10, 13, 44, 109
coding, viii, 118
complement, 6, 10, 14
complex numbers, 65
computation, 156, 175, 176
concreteness, 161
configuration, 42
conjugation, 163, 166, 167, 174

construction, 95, 97, 104, 106, 111
continued fractions, 128, 139
convergence, 17, 44, 56
correlation, 143, 165
correlations, 150
cuprates, 150

D

decoding, 155, 156, 166, 170, 172
decomposition, 6, 82, 90, 104
degenerate, 135
DEL, 19, 24, 26
demonstrations, 1
density matrices, 170
differential equations, vii, viii, 19, 20, 21, 37, 41, 42, 45, 46, 47, 48, 58, 124, 126, 129, 132, 137, 139, 140
diffusion, viii, 46
Dirac equation, 168
duality, 11, 12, 16

E

cconomics, 9
electromagnetic, 124, 125
electromagnetic fields, 124
electron, 142, 143, 144, 145, 146, 147, 150
energy, 123, 124, 128, 135, 150, 154
engineering, vii, viii, 41, 42
enlargement, 93, 94, 95, 97, 99, 100
equality, 3, 23, 73

equilibrium, 15
Euclidean space, 29
evolution, viii, 41, 42, 43, 44, 47, 48, 51, 54, 58, 59, 159, 160, 161, 162, 163, 166, 169, 172

F

families, 81, 111, 113
fermions, 150
ferromagnetism, 143
fiber bundles, 120
fluctuations, 144
formation, 143
formula, 67, 68, 76, 78, 83, 87, 90, 91, 92, 94, 97, 98, 101, 102, 131
foundations, 140, 153

G

geometry, 48, 62, 120

H

Hamiltonian, ix, 124, 125, 130, 131, 135, 137, 141, 142, 143, 144, 148, 149, 156, 157, 159, 160
Hermitian operator, 168
Hilbert space, vii, viii, ix, 1, 2, 3, 6, 7, 8, 9, 10, 11, 15, 16, 17, 19, 20, 21, 32, 38, 39, 61, 62, 63, 69, 88, 89, 111, 113, 118, 120, 123, 130, 131, 136, 140, 141, 144, 145, 151, 153, 154, 155, 156, 157, 158, 170, 171, 173, 174, 176
Hubbard model, ix, 143, 144, 145, 146, 147, 149, 150, 151

I

inequality, 3, 8, 11, 24, 26, 30, 32, 33, 36, 44, 50, 54, 55, 68
initial state, 42, 164, 169
involution, 63, 65, 80, 81
ions, 174, 175, 176

L

Lie algebra, 102, 119, 130, 136
Lie group, 62
light, viii, 124, 140
linear function, vii, 8, 9, 90, 154
linear systems, 43, 47
local conditions, 42, 43
logarithmic functions, 107

M

magnetic field, 162
mapping, 47, 51, 67, 69, 71, 74, 81, 86, 87, 91, 92, 93, 94, 95, 98, 100, 103, 106, 115, 154, 155, 156, 158, 163, 164, 165, 166, 167, 170, 171, 172, 173
master equation, 159
materials, viii, 42
mathematics, 42, 48, 52, 61, 154
matrix, ix, 30, 31, 125, 132, 137, 143, 144, 148, 149, 156, 157, 158, 164, 167
matter, viii, 124, 140, 143
mean-field theory, 143, 144
measurement, 46, 122, 163, 168, 169, 170, 171, 172, 177, 178
measurements, 42, 171, 176, 177
models, viii, 123, 124, 125, 139, 140, 141, 146, 149, 172, 175
modules, viii, 61, 63, 82, 83, 84, 85, 93, 94, 95, 96, 97, 100, 103, 113, 118, 121
modulus, 49
multiplication, 65, 67, 69, 74, 76, 77, 79, 81, 87, 91, 95, 96, 97, 100, 101, 103, 108, 112

N

Nash equilibrium, 18
noncommutativity, 63
nonlinear systems, 43, 58
nuclear spins, 177

O

one dimension, 143
operations, 9, 108, 154, 174

operations research, 9
optimization, 1, 2, 3, 16, 17
ordinary differential equations, 43
orthogonality, vii, 1, 2

P

partial differential equations, 41, 42, 47
particle physics, 119
photonics, 159, 168
photons, 125
physical laws, 177
physical phenomena, 42, 168
physical properties, 173
physics, viii, ix, 46, 62, 124, 140, 141, 143, 153, 154, 168, 173, 174, 176
Planck constant, 156
programming, 1, 9, 10, 12, 16, 17

Q

quantum bits, ix, 153, 154
quantum computing, 165
quantum dot, 125, 140, 142
quantum electrodynamics, viii, 124
quantum field theory, 62
quantum mechanics, 142, 154, 155, 166, 168, 173, 178
quantum optics, viii, 124, 140
quantum state, ix, 153, 154, 155, 158, 163, 164, 165, 166, 167, 168, 169, 171, 173, 178
qubits, 160, 161, 162, 165, 177

R

Rabi models, 123, 125, 139, 141
recurrence, 87, 124, 127, 129, 133, 134, 138, 139, 142
reflectivity, 140
relativity, 142
repulsion, 144, 148
restrictions, 10, 11, 74, 90, 106

S

semiconductor, viii, 124, 140

semigroup, 19, 20, 43, 44, 46, 47, 51
simulation, ix, 153, 172, 173, 174, 175, 176
simulations, 172, 175
solution, 11, 21, 23, 24, 26, 46, 47, 51, 52, 53, 57, 117, 124, 128, 134, 138, 139, 151, 157
spacetime, 163, 164
special relativity, 164
spin, 124, 144, 145, 146, 147, 148, 160, 178
square lattice, 145
stability, vii, viii, 19, 20, 21, 25, 38, 39
stabilization, 52
statistics, 165, 176
structure, viii, 16, 62, 63, 69, 79, 83, 88, 94, 95, 96, 98, 99, 100, 111, 113, 118, 120, 133, 141
superconductivity, 144, 150
symmetry, viii, ix, 123, 124
synthesis, 38

T

techniques, 123, 129, 143
technologies, 42, 173
thermalization, 125, 141
topology, 14, 44, 70, 72, 88, 104, 105, 107, 121
transformation, ix, 11, 13, 125, 130, 135, 153, 154, 163, 164, 165, 168
transformations, viii, 100, 123, 124, 140
transition metal, 143
transmission, 122

V

variables, 12, 15, 120
variations, 125
vector, ix, 2, 6, 9, 10, 11, 14, 17, 75, 77, 121, 153, 154, 163, 169, 170
velocity, 42

Y

yin, 13

Related Nova Publications

NEXT GENERATION NEWTON-TYPE METHODS

AUTHOR: Ram U. Verma

SERIES: Mathematics Research Developments

BOOK DESCRIPTION: This monograph is aimed at presenting "Next Generation Newton-Type Methods," which outperform most of the iterative methods and offer great research potential for new advanced research on iterative computational methods.

HARDCOVER ISBN: 978-1-53615-456-6
RETAIL PRICE: $160

NEW TRENDS IN FRACTIONAL PROGRAMMING

AUTHOR: Ram U. Verma

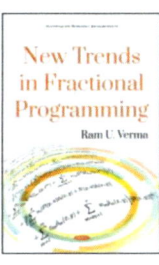

SERIES: Mathematics Research Developments

BOOK DESCRIPTION: This monograph presents smooth, unified, and generalized fractional programming problems, particularly advanced duality models for discrete min-max fractional programming.

HARDCOVER ISBN: 978-1-53615-371-2
RETAIL PRICE: $230

To see a complete list of Nova publications, please visit our website at www.novapublishers.com

Related Nova Publications

MATHEMATICAL MODELING OF REAL WORLD PROBLEMS: INTERDISCIPLINARY STUDIES IN APPLIED MATHEMATICS

EDITORS: Zafer Aslan, Funda Dökmen, Enrico Feoli, and Abul H. Siddiqi

SERIES: Mathematics Research Developments

BOOK DESCRIPTION: Data mining provides avenues for proper understanding of real world problems. For researchers interested in data mining and new applications, this book is a multidisciplinary 'handbook' in data processes, engineering and medical applications.

HARDCOVER ISBN: 978-1-53616-267-7
RETAIL PRICE: $230

MATHEMATICAL MODELING FOR THE SOLUTION OF EQUATIONS AND SYSTEMS OF EQUATIONS WITH APPLICATIONS. VOLUME III

AUTHORS: Ioannis K. Argyros and Santhosh George

SERIES: Mathematics Research Developments

BOOK DESCRIPTION: These books contain a plethora of updated bibliography and provide comparison between various investigations made in recent years in the field of computational mathematics in the wide sense.

HARDCOVER ISBN: 978-1-53615-942-4
RETAIL PRICE: $270

To see a complete list of Nova publications, please visit our website at www.novapublishers.com